Plant
Systematics
and
Evolution   Supplementum 4

T. J. Mabry · G. Wagenitz (eds.)
Research Advances in the *Compositae*

Springer-Verlag Wien New York

Prof. T. J. Mabry
Department of Botany
The University of Texas at Austin
Austin, Texas, U.S.A.

Prof. G. Wagenitz
Systematisch-Geobotanisches Institut
der Universität Göttingen
Göttingen, Federal Republic of Germany

Softcover reprint of the hardcover 1st edition 1990

Printed on acid-free paper

With 20 Figures

ISSN 0172-6668
ISBN-13: 978-3-7091-7437-1   e-ISBN-13: 978-3-7091-6928-5
DOI: 10.1007/978-3-7091-6928-5

Contents

Pl. Syst. Evol. [Suppl. 4], 1 – 2 (1990)

Introduction

T. J. MABRY and G. WAGENITZ

The half-day symposium on "Multidisciplinary approaches to the systematics of *Compositae*", held as part of the XIV International Botanical Congress in Berlin, on July 26, 1987, was designed to complement the University of Reading *Compositae* Conference (1975). The latter had yielded two impressive and thorough volumes on "The biology and chemistry of the *Compositae*", which were edited by HEYWOOD, HARBORNE & TURNER (1977). The 1987 Berlin Symposium did not attempt to update the information from the earlier conference but instead focussed on selected new methods for investigating the systematics of the family as well as a few examples of new systematic approaches with classical methods.

From mapping chloroplast DNA restriction sites JANSEN, PALMER, and MICHAELS reported the astonishing fact that, with the exception of one group (the subtribe *Barnadesiinae* of the tribe *Mutisieae*), all investigated other members of *Compositae* exhibit a characteristic inversion in their chloroplast DNA, suggesting that the inversion occurred early in the evolution of the family and that at least its major part is monophyletic. Within those groups with the inverted segment, chloroplast DNA also suggests that most of the conventionally recognized tribes are also monophyletic. This lends high credit to our predecessors who laid the foundations for the taxonomic system of the *Compositae*. These chloroplast DNA studies have already been published and are not included here (JANSEN & PALMER 1987, 1988). Another report concerning the whole family was given by BREMER on the use of cladistic methods for analyzing morphological and phytochemical data for establishing the affinities of the *Compositae* tribes; these data presented in another symposium of the congress also have been published elsewhere (BREMER 1987). Together, these published investigations as well as the studies presented here will surely provoke further discussions on classifications of the family and stimulate further investigations.

While phytoserology has been used in systematics for several decades, the methods have been markedly refined in recent times. The usefulness for systematics of serological studies of the main seed storage protein of members of the *Compositae* is clear from the paper by FISCHER & JENSEN. How important patterns of secondary plant constituents are for understanding phylogenetic relationships is well illustrated by the outstanding chemosystematic contribution by BOHLMANN and coworkers for tribes *Vernonieae* and *Heliantheae*.

During the last decade cladistics has emerged as a major tool of systematics for the classificatory process. However, while the rigor of this approach appeals to many taxonomists, its successful application remains, in many cases, elusive. Such difficulties are evident from the paper by KEELEY & TURNER in which they attempt to elucidate by cladistic analysis the systematics of the large genus *Vernonia*. They emphasize that the direction of character evolution is questionable and that parallelism appears to occur over and over again, thus leaving many phylogenetic questions unsettled.

Experimental taxonomy including field studies, karyological investigations and artificial hybridization have been applied to many genera in both, N. America and Europe, but much less is known about taxa from areas outside these continents. In a careful synthesis of all data available world-wide for *Heliantheae* SUNDBERG & STUESSY report that reproductive isolation is the dominant isolating factor, with geographical isolation occurring in about two thirds of the taxa.

The importance of the classical approach to systematics is illustrated by BARKLEY with his studies on members of the genus *Senecio* from Mexico and C. America.

Finally, we wish to thank all the speakers for their contributions to the symposium. We are also indebted to Prof. EHRENDORFER and Springer-Verlag, Wien, who enabled publication of the symposium as a supplement issue of *Plant Systematics and Evolution*.

References

BREMER, K., 1987: Tribal interrelationships of the *Asteraceae*. — Cladistics 3: 210—253.

HEYWOOD, V. H., HARBORNE, J. B., TURNER, B. L., (Eds.) 1977: The biology and chemistry of the *Compositae*. 2 vols. — London: Academic Press.

JANSEN, R. K., PALMER, J. D., 1987: A chloroplast DNA inversion marks an ancient evolutionary split in the sunflower family (*Asteraceae*). — Proc. Natl. Acad. Sci. USA 84: 5818—5822.

— — 1988: Phylogenetic implications of chloroplast DNA restriction site variation in the *Mutisieae*. — Am. J. Bot. 75: 753—766.

Pl. Syst. Evol. [Suppl. 4], 3–43 (1990)

Progress in the chemistry of the *Vernonieae* (*Compositae*)

F. Bohlmann and J. Jakupovic

Received December 4, 1987

Key words: Angiosperms, *Compositae, Vernonieae*. – Phytochemistry, germacranolides, glaucolides, 5-alkylcoumarins, sesquiterpenes.

Abstract: The *Vernonieae* are characterized by the occurrence of highly oxygenated germacranolides and by simple polyacetylenes as well as by the absence of diterpenes, phenyl propanes and p-hydroxyacetophenone derivatives. However, some groups of genera or species differ from this general picture. Glaucolides are absent, with two exceptions, from species of the subtribes *Pseudostifftinae, Lychnophorinae, Centratherinae* and *Elephantopinae*, but common in the *Vernoniinae*, especially in the large genus *Vernonia*, but also in the subtribes *Piptocarphinae* and *Rolandrinae*. Again there are exceptions in the subtribe *Vernoniinae*. Some genera and a few *Vernonia* spp. only contain the rare 5-alkylcoumarins, guaianolides, elemanolides or furoheliangolides. The chemistry of the African genus *Corymbium* supports its exclusion from the tribe.

Ten years ago, the chemistry of the tribe *Vernonieae* was reviewed by Harborne & Williams (1977) and they ended with the conclusion: "so little is known of the chemistry of the *Vernonieae* that it is not possible to draw any firm conclusions about the utility of secondary compounds as systematic markers in the tribe". This was surely true, especially as the taxonomy of the *Vernonieae* also still was poorly understood. Even now no comprehensive modern treatment of the whole group is available. However, our knowledge of the chemistry has increased. While 10 years ago only some results from nine of the 70 genera were available, now about 240 species out of 30 genera have been investigated (Table 3).

As already pointed out previously flavones and flavonols are widespread in the tribe. They have been detected in nearly all species investigated. As only very few recent results are available in this field the tables of Harborne & Williams (1977) are still valid. These authors also have already mentioned the widespread occurrence of sesquiterpene lactones, most of them belonging to germacranolides of the glaucolide-type, but also other highly oxygenated germacranolides and elemanolides as well as simple guaianolides. In addition to the occurrence of simple polyacetylenes, like tridecapentaynene and the related thiophenes, the isolation of triterpenes (Das & al. 1983) and some less widespread compounds were reported, e.g., vernolic acid, present in the seed oil of some *Vernonia* species.

In the meantime, a lot of new results on the chemistry of the *Vernonieae* has become available. Especially, the number of isolated sesquiterpene lactones and

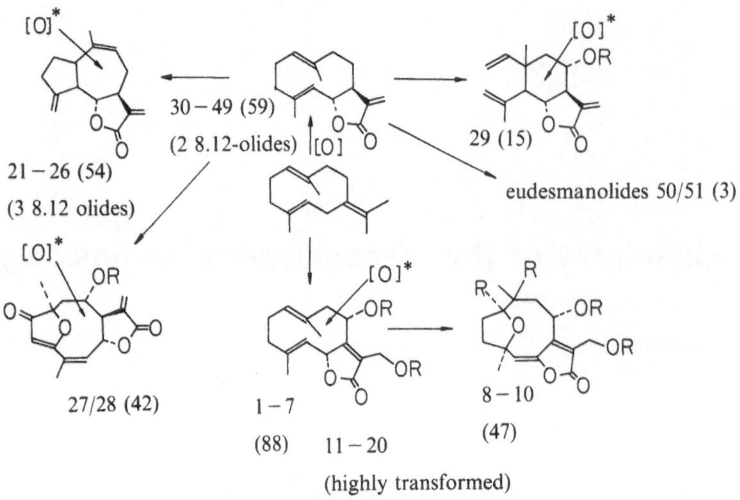

Fig. 1. Main pathways of sesquiterpenes in the *Vernonieae* (number of compounds in parenthesis). *Indicates further oxidation

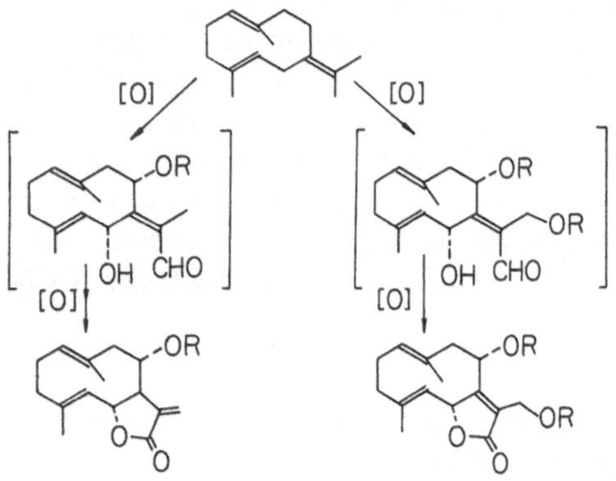

Fig. 2. Proposed oxidative transformations of germacrene B

their derivatives has been raised drastically. The total number of lactones now has increased to more than 340 belonging to about 20 different types as shown in Tables 1 and 2. Their distribution among *Vernonieae* is listed in Table 3 [genera arranged following the proposal of ROBINSON & al. (1980) and for *Vernonia* according to the subdivision of JONES (1981 a, b)].

Most sesquiterpene lactones are glaucolides and related hirsutinolides. As shown in Fig. 1 the main biogenetic pathways in the tribe can be characterized by an unusual high degree of enzymatic oxidation of the simple precursor. As already assumed by FISCHER & al. (1979), germacrene B may be the common precursor for all sesquiterpene lactones which are formed by allylic oxidation at C-6, C-8 and C-12. A typical difference in parts of the *Vernonieae* apparently is the ability of an additional allylic oxidation at C-13 leading to 1 as shown in Fig. 2. So far, there are no experimental data available supporting this assumption. However, the fact that about 90% of all sesquiterpene lactones are 6,12-olides suggests oxidation at the most preferred position of germacrene B. Further transformation of 1 then would lead to glaucolides of types 4 and 5 which surely are the precursors of the

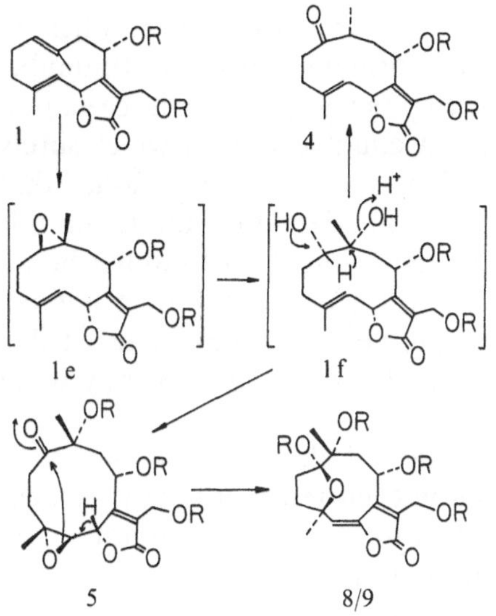

Fig. 3. Formation of glaucolides and hirsutinolides

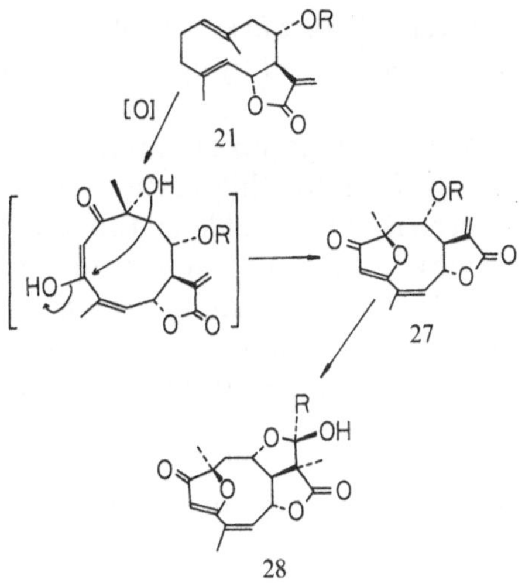

Fig. 4. Formation of furoheliangolides

hirsutinolides of type 8/9 (Fig. 3). The direct precursor of 4 and 5, the epoxide 1 e, still has not been reported. Most likely, 1 e first is hydrolyzed to 1 f which then either is oxygenated to 5 or rearranged to lactones of type 4. The formation of the hirsutinolides (8/9) requires opening of the 4,5-epoxide. The resulting carbinol then would directly be transformed to the semiacetal function which is typical for this type of lactones. Though most glaucolides and hirsutinolides with no oxygen function at C-10 have, as expected, a 10 α-methyl group, some are known with the β-configuration. Most likely, this is due to subsequent isomerization which seems to be favoured by the hydroxy groups at C-9 or C-3 which may change the preferred conformation.

In connection with the absolute configuration of the glaucolides, which can not be determined easily by circular dichroism, lactone 15 is important. In this case, the observed Cotton-effect could be used directly to establish the absolute configuration. As the lactone 15 is biogenetically derived from a typical glaucolide, most likely all glaucolides have the same absolute configuration.

Another, relatively large group of lactones in the *Vernonieae*, the furohelian-golides of type 27/28, probably are formed by transformation of the 1(10)-double bond, as shown in Fig. 3 for the glaucolides, followed by subsequent allylic oxidation at C-3 (Fig. 4). The resulting 1,3-diketone then easily could lead to 27 which surely also is the precursor of the usual goyazensolides (28). It should be mentioned that allylic oxidation at C-3 obviously causes a cis-/trans-isomerization of the 4,5-double bond. Similar oxidation at C-2 or C-14 leads to isomerization of 1(10)-double bond.

A wide range of further common or in part rare transformations are observed. They give rise to the enormous variety of sesquiterpene lactones in this tribe. Probably these unusual lactones like 6, 7, 11 − 19, 23, 24, 25, and 29 may be in part of chemotaxonomic relevance.

The overall picture of the sesquiterpene lactones of the tribe *Vernonieae* is relatively uniform and characterized by the common occurrence of highly oxygen-ated germacranolides. As shown in Figs. 2 − 4 even the biogenetic variations are limited. Thus, the formation of furoheliangolides required several identical steps as in the case of glaucolides. Even the elemanolides like vernolepin obviously are derived from highly oxygenated germacranolides. The cooccurrence of the related lactones vernolide (25 a) and vernolepin (29 g) in several *Vernonia* spp. is of interest as in the case of vernolide the elemanolide formation is blocked by epoxidation of the Δ1(10) double bond.

Some species contain the very rare 5-alkyl coumarins, a group of natural com-pounds most likely derived from polyketides and not from the shikimic acid path-way. Up to now, 40 compounds of this type have been isolated (Table 1). Especially rich is *Ethulia conyzoides* which contains ten of these compounds. But these cou-marins were also reported from *Bothriocline, Erlangea, Volkensia*, and two *Vernonia* spp., which lack the usually present sesquiterpene lactones. It should be noted that similar compounds only have been isolated from representatives of the tribe *Mu-tisieae,* subtribe *Mutisiinae.*

Whereas sesquiterpene lactones are very widespread in the whole tribe, other sesquiterpenes, except common hydroxycarbons, have not been reported very often. The nerolidol derivatives 77 and 78 (Table 3) have been isolated from *Lychnophora ericoides,* from *Pseudostifftia kingii* the diol 80, from *Vernonia fastigiata* the triol 79 and from three other *Vernonia* spp. the furane derivatives 28 and 83 as well as the precursor 85 and the degradated derivative 84.

In *Lychnophora sellowii* the bisabolane derivatives 86 a − 86 c are present. The latter is also reported from *Vernonia galpinii* and *V. hirsuta.* From *V. neocorymbosa* the ketones 86 d and 86 c were isolated, while *V. echitifolia* contains 87. Two further derivatives (88 a and 89) are isolated from a *Pseudoelephantopus* species and *Ver-nonia aurea,* respectively. From two *Elephantopus* spp. the senecioate 88 b is re-ported.

A few eudesmane derivatives are also present in members of the tribe. Thus *Alcantara ekmaniana* afforded the acid 90 a − 90 c, 91 a, and 91 c and from *Vernonia* spp. the compounds 91 b, 93 a, and 93 b as well as 92 a − 92 c are isolated. An other *Vernonia* species afforded the isomeric aldehydes 95 a and 95 b.

Unique sesquiterpenes are isolates from *Lychnophora* species. In addition to the caryophyllene derivatives 96 a − 96 c the α-humulene derivatives 99 a − 99 c and 100 a − 100 c are reported. The very unique acid 102 most likely is derived from 99 c.

Some miscellaneous compounds are isolated from single species. Thus, from *Lychnophora passerina* the cyclopropenone 106, from *Vernonia sutherlandia* the acetates 105 and 107, from *V. cotoneaster* the spathulenol derivatives 103 a and 103 b, and from *V. adonensis* the highly oxygenated thymol derivative 107.

Characteristic for the whole tribe is the absence of diterpenes, phenylpropanes (except *V. echitifolia* with lignanes) and p-hydroxyacetophenone derivatives, otherwise all are wide-spread in the *Compositae*.

All these data allow a clear separation of the tribe *Vernonieae* from the other tribes. However, a systematic separation of the genera by their chemistry is only partly possible. As shown in Tables 3 and 4 glaucolides and related lactones are not only reported from the subtribe *Vernoniinae* but also from the *Piptocarphinae* and *Rolandrinae* as well as from the genus *Chresta* which has been placed in the *Lychnophorinae*, where all other genera contain furoheliangolides. Furthermore, glaucolides are present in *Pseudoelephantopus* while other members of the *Elephantopinae* have unique germacranolides like elephantopin.

Especially complicated is the situation in the subtribe, *Vernoniinae*, containing most of the genera (Table 5). The chemistry of many of these differs from that of the main genus *Vernonia*. Thus, in *Bothriocline, Ethulia, Volkensia*, and parts of *Erlangea*, sesquiterpene lactones are absent and replaced by the rare 5-alkyl coumarins. *Gutenbergia, Heterocoma, Hoplophyllum*, and parts of *Erlangea* only have guaianolides, while *Alcantara* and *Oliganthes* afford furoheliangolides. *Mattfeldanthus* contains 8,12-germacranolides. *Gongrothamnus* has simple elemanolides and differs from *Vernonia* by yellow flowers. Therefore, these genera should be better separated from *Vernonia*. *Corymbium* must be excluded from the tribe as the sesquiterpene lactones are here replaced by diterpenes (108 − 112). The presence of these labdanes together with tridecapentaynene may be an indication that *Corymbium* should be better placed in the *Eupatorieae* or *Inuleae*. Perhaps a further subdivision of the *Vernoniinae* is necessary.

Table 1. Sesquiterpene lactones from *Vernonieae*. Abbreviations see Table 2

		R	R'	X	
	1a	H	H	β-OAc, H	Jakupovic & al. (1986 d)
	1b	Ac	Ac	β-OAc, H	Bohlmann & Zdero (1982 f), Bohlmann & al. (1982 g)
	1c	Ac	Meacr	α-OAc, H	Bohlmann & Zdero (1982 f)
	1d	Ac	Meacr	O	Bohlmann & Zdero (1982 f)

		R	R'	X	
	2a	Ac	Ac	H_2	Jakupovic & al. (1986 d)
	2b	Ac	Meacr	H_2	Bohlmann & al. (1981 g)
	2c	Ac	Ang	H_2	Betkoushi & al. (1976)
	2d	Ac	Tigl	H_2	Padolina & al. (1974 a)
	2e	Ac	Ac	β-OAc, H	Bohlmann & al. (1981 c), Bohlmann & Zdero (1982 f)
	2f	Ac	Meacr	α-OAc, H	Betkoushi & al. (1975)
	2g	Ac	Epmeacr	α-OAc, H	Betkoushi & al. (1975)
	2h	Ac	Epmeacr	O	Bohlmann & al. (1984 b)
	2i	Ac	Ac	α-OAc, H	Jakupovic & al. (1987 c)
	2j	Ac	Hymeacr	H	Jakupovic & al. (1986 a)
	2k	H	Tigl	H_2	Jakupovic & al. (1986 c)

		R	R'	
	3a	Ac	Ac	Catalan & al. (1985), Jakupovic & al. (1986 d)
	3b	iBu	Ac	Jakupovic & al. (1986 d)
	3c	Meacr	Ac	Bohlmann & Zdero (1982 f), Zabel & al. (1980)
	3d	Epmeacr	Ac	Bohlmann & al. (1984 b)
	3e	Ang	Ac	Jakupovic & al. (1986 d)
	3f	Hymeacr	Ac	Jakupovic & al. (1987 c)
	3g	D	Ac	Jakupovic & al. (1986 g)
	3h	Tigl	Ac	Jakupovic & al. (1986 c)
	3i	Tigl	H	Jakupovic & al. (1986 c)

		R	R'	R"	R'''	
	4a	H	Meacr	H	OH	Jakupovic & al. (1986 d), Bohlmann & al. (1982 h)
	4b	Ac	Meacr	H	OAc	Jakupovic & al. (1986 d), Bohlmann & al. (1982 h)
	4c	Ac	Tigl	H	OAc	Bohlmann & al. (1981 f), Jakupovic & al. (1986 b)
	4d	Ac	H	OH	H	
	4e	Ac	Meacr	OH	H	Jakupovic & al. (1987 c)
	4f	Ac	Tigl	OH	H	Jakupovic & al. (1987 c)
	4g					Jakupovic & al. (1987 c)

Table 1 continued

	R	X	R'	R''	
5a	Meacr	α-OH, H	Me	H	Jakupovic & al. (1986d), Bohlmann & al. (1982h)
5b	Tigl	α-OH, H	Me	H	Jakupovic & al. (1986d)
5c	Ac	α-OAc, H	Me	H	Toubiana & al. (1972)
5d	Meacr	α-OAc, H	Me	H	Jakupovic & al. (1986d), Bohlmann & al. (1982h)
5e	Tigl	α-OAc, H	Me	H	Jakupovic & al. (1986d), Bohlmann & al. (1982h)
5f	Meacr	O	Me	H	Jakupovic & al. (1986d), Bohlmann & al. (1982h)
5g	Hymeacr	O	Me	H	Bohlmann & al. (1984f)
5h	Ang	O	Me	H	Jakupovic & al. (1986d)
5i	Tigl	O	Me	H	Jakupovic & al. (1986d), Bohlmann & al. (1982h)
5j	Ac	O	Me	H	
5k	Ac	O	H	Me	Catalan (1985)
5l	Ac	O	OAc	Me	Padolina & al. (1974b)
5m	Pro	O	OAc	Me	Bohlmann & al. (1981c)
5n	Meacr	O	OAc	Me	Cox & Sim (1975), Padolina & al. (1974b)
5o	Hymeacr	O	OAc	Me	Bohlmann & Czerson (1978c)
5p	Ang	O	OAc	Me	Jakupovic & al. (1986b)
5q	Tigl	O	OAc	Me	Dominguez & al. (1986)
5r	Sen	O	OAc	Me	Jakupovic & al. (1986b)

	X	R	R'	
5s	β-Me, H	OH	Meacr	Jakupovic & al. (1987c)
5t	CH₂	H	Tigl	Jakupovic & al. (1986c)

	X	
5u	β-OH, H	Bohlmann & al. (1982d)
5v	O	Jakupovic & al. (1986a)

	R	
5w	H	Jakupovic & al. (1986c)
5x	Ac	Jakupovic & al. (1986c)

5y		Jakupovic & al. (1986a)

Table 1 continued

		R	R′	R″	
	6a	Meacr	H	Ac	Jakupovic & al. (1987 c)
	6b	Ang	H	Sen	Jakupovic & al. (1987 c)
	6c	Ang	H	4 Hysen	Jakupovic & al. (1987 c)
	6d	Meacr	OAc	Ac	Jakupovic & al. (1987 c)
	6e	Meacr	OAc	Sen	Jakupovic & al. (1987 c)
	6f	Meacr	OAc	4 Hysen	Jakupovic & al. (1987 c)
	6g	Epmeacr	OAc	Ac	Jakupovic & al. (1987 c)
	6h	Epmeacr	OAc	Sen	Jakupovic & al. (1987 c)
	6i	Epmeacr	OAc	4 Hysen	Jakupovic & al. (1987 c)
	6j	Ang	OAc	Ac	Jakupovic & al. (1987 c)
	6k	Ang	OAc	Sen	Jakupovic & al. (1987 c)
	6l	Ang	OAc	4 Hysen	Jakupovic & al. (1987 c)
	6m	Tigl	OAc	Ac	Jakupovic & al. (1987 c)

	6n		Jakupovic & al. (1987 c)

		R	R′	
	6o	Meacr	H	Bohlmann & al. (1984 b)
	6p	Mebu	H	Bohlmann & al. (1984 b)
	6q	Ang	Ac	Jakupovic & al. (1987 c)

		X	
	6r	CHCH$_3$	Jakupovic & al. (1987 c)
	6s	β-CH$_2$CH$_3$, H	Jakupovic & al. (1987 c)

		R	
	7a	H	Herz & al. (1981)
	7b	Ac	Herz & al. (1981)

	7c	Jakupovic & al. (1987 c)

		R	R′	
	7d	H	Capr	Bohlmann & al. (1982 g), Jakupovic & al. (1986 d)
	7e	Ac	Meacr	Jakupovic & al. (1986 d)
	7f	Ac	Ang	Jakupovic & al. (1986 d)

Table 1 continued

	R					
7g	Ang					JAKUPOVIC & al. (1986d)
7h	Meacr					JAKUPOVIC & al. (1986a)
7i						JAKUPOVIC & al. (1986d)

	R	R′	R″	R‴	R⁗	
8a	H	Capr	Me	OH	H	JAKUPOVIC & al. (1986d), BOHLMANN & al. (1982g)
8b	Ac	Ang	Me	OH	H	JAKUPOVIC & al. (1986d)
8c	Ac	Capr	Me	OH	H	JAKUPOVIC & al. (1986a), BOHLMANN & al. (1982g)
8d	Ac	Epmeacr	Me	OH	H	JAKUPOVIC & al. (1987c)
8e	H	Meacr	H	Me	OH	JAKUPOVIC & al. (1986d), BOHLMANN & al. (1978a)
8f	H	Epmeacr	H	Me	OH	JAKUPOVIC & al. (1986d), BOHLMANN & al. (1978a)
8g	Ac	Meacr	H	Me	OH	JAKUPOVIC & al. (1986d), BOHLMANN & al. (1978a)
8h	Ac	Epmeacr	H	Me	OH	JAKUPOVIC & al. (1986d), BOHLMANN & al. (1978a)
8i	Ac	Hymeacr	H	Me	OH	JAKUPOVIC & al. (1986d), BOHLMANN & al. (1978a)
8j	Ac	Ang	H	Me	OH	JAKUPOVIC & al. (1986d)
8k	Ac	Tigl	H	Me	OH	JAKUPOVIC & al. (1986d), BOHLMANN & al. (1982h)
8l	Ac	4Hytigl	H	Me	OH	JAKUPOVIC & al. (1986a)
8m	Ac	Meacr	H	CH$_2$OH	OH	JAKUPOVIC & al. (1986d), BOHLMANN & al. (1978a)
8n	Et	Ac	Me	H	OH	HOLUB (pers. comm.)

	R	R′	R″	R‴	
9a	H	Ac	H	H	JAKUPOVIC & al. (1986d)
9b	H	Meacr	H	H	JAKUPOVIC & al. (1986d)
9c	Me	Ac	H	H	JAKUPOVIC & al. (1986d)
9d	Me	Meacr	H	H	JAKUPOVIC & al. (1986d)
9e	Me	Ac	H	Me	JAKUPOVIC & al. (1986d)
9f	Me	Meacr	H	Me	JAKUPOVIC & al. (1986d)
9g	Ac	H	H	H	JAKUPOVIC & al. (1986d), COWALL & al. (1981)
9h	Ac	Ac	H	H	JAKUPOVIC & al. (1986d), BOHLMANN & al. (1979)
9i	Ac	Pro	H	H	JAKUPOVIC & al. (1986d), BOHLMANN & al. (1979)

Table 1 continued

	R	R′	R″	R‴	
9j	Ac	Meacr	H	H	Jakupovic & al. (1986d), Cowall & al. (1981)
9k	Ac	Tigl	H	H	Jakupovic & al. (1986d), Cowall & al. (1981)
9l	Ac	Ac	H	iVa	Jakupovic & al. (1986d), Bohlmann & al. (1981c)
9m	Ac	Meacr	H	Et	Jakupovic & al. (1986d), Cowall & al. (1981)
9n	Ac	Pro	Me	H	Jakupovic & al. (1986d), Bohlmann & al. (1979)
9o	Ac	Tigl	Me	H	Jakupovic & al. (1986d), Bohlmann & al. (1981f)
9p	Ac	Ac	Ac	H	Jakupovic & al. (1986d), Bohlmann & al. (1979)
9q	Ac	Pro	Ac	H	Jakupovic & al. (1986d), Bohlmann & al. (1979)
9r	Ac	Ac	Ac	Me	Jakupovic & al. (1986d), Bohlmann & al. (1979a)
9s	Ac	Meacr	Ac	Me	Jakupovic & al. (1986d), Bohlmann & al. (1983b)
9t	Et	Meacr	H	H	Jakupovic & al. (1986d), Cowall & al. (1981)
9u	Meacr	H	H	H	Jakupovic & al. (1986d), Herz & Kulanthaivel (1983)
9v	Tigl	H	H	H	Jakupovic & al. (1986d), Herz & Kulanthaivel (1983)
9w	Ac	Hymeacr	H	H	Jakupovic & al. (1986a)
9x	Ac	4Hytigl	H	H	Jakupovic & al. (1986a)
9y	Et	Ac	H	H	Catalan & al. (1986)

	R	R′	R″	X	
10a	Ac	Ac	OH	H₂	Jakupovic & al. (1986a), Bohlmann & al. (1981e)
10b	Ac	Meacr	OH	H₂	Jakupovic & al. (1986a), Bohlmann & al. (1978a)
10c	H	4-Hysen	H	O	Bohlmann & al. (1983a)
10c	H	4-Hysen	H	O	Bohlmann & al. (1983a)
10d	Ac	Meacr	H	O	Bohlmann & al. (1983a)
10e	Ac	Sen	H	O	Bohlmann & al. (1983a)
10f	Ac	4-Hysen	H	O	Bohlmann & al. (1983a)
10g	Ac	5-Hysen	H	O	Bohlmann & al. (1983a)
10h	Ac	5-Acsen	H	O	Bohlmann & al. (1983a)

10i		Jakupovic & al. (1986a)

Table 1 continued

		R	R′	
	11a	H	H	JAKUPOVIC & al. (1986d)
	11b	Ac	H	JAKUPOVIC & al. (1986d)
	11c	Ac	Ac	JAKUPOVIC & al. (1986d), BOHLMANN & al. (1980a)
	11d			BOHLMANN & al. (1981c), JAKUPOVIC & al. (1986d)
	11e			BOHLMANN & al. (1981c), JAKUPOVIC & al. (1986d)

		R	R′	
	12a	H	Ang	JAKUPOVIC & al. (1986d)
	12b	Me	Ac	JAKUPOVIC & al. (1986d)
	12c	Me	Ang	JAKUPOVIC & al. (1986d)
	12d	Ac	Ac	JAKUPOVIC & al. (1986d)
	12e	Ac	Ang	JAKUPOVIC & al. (1986d)
	12f	H	Tigl	JAKUPOVIC & al. (1986c)
	12g			JAKUPOVIC & al. (1986c)
	12h			JAKUPOVIC & al. (1986d)

		R	R′	
	13a	H	Ac	JAKUPOVIC & al. (1986d)
	13b	Me	Ac	JAKUPOVIC & al. (1986d)
	13c	Me	iBu	JAKUPOVIC & al. (1986d)
	13d	Me	Meacr	JAKUPOVIC & al. (1986d)

		R	
	14a	Ac	JAKUPOVIC & al. (1986d), BOHLMANN & al. (1981g)
	14b	Meacr	JAKUPOVIC & al. (1986d), BOHLMANN & al. (1981g)
	14c	Tigl	JAKUPOVIC & al. (1986d), BOHLMANN & al. (1981g)

Table 1 continued

		R			
	15				Warning & al. (1987)
	16				Bohlmann & al. (1981 g)

		R			
	17a	Meacr			Bohlmann & al. (1981 c)
	17b	Tigl			Bohlmann & al. (1981 c)

		R			
	18a	Meacr			Jakupovic & al. (1986 d)
	18b	Tigl			Jakupovic & al. (1986 c)

	19a				Jakupovic & al. (1986 c)

		R			
	19b	H			Jakupovic & al. (1986 c)
	19c	Tigl			Jakupovic & al. (1986 c)

	19d				Jakupovic & al. (1986 c)

	20				Jakupovic & al. (1986 c)

		R	R′	R″	
	21a	H	Me	Me	
	21b	H	CH₂OSen	Me	Bohlmann & al. (1978 a)
	21c	H	CH₂OiVal	Me	Bohlmann & al. (1978 a)
	21d	OSen	Me	Me	Bohlmann & Zdero (1977 a)
	21e	OMeacr	CH₂OH	CH₂OH	Mompon & al. (1973)
	21f	OMeacr	CH₂OH	CHO	Mompon & Toubiana (1976)
	21g	H	COOH	Me	Banerjee & al. (1986)
	21h	OHymeacr	CH₂OH	CH₂OH	Jakupovic & al. (1986 b)

Table 1 continued

		R	R′		
	22a	H	H		
	22b	OMeacr	OH		JAKUPOVIC & al. (1987 b)

		R		
	22c	H		GERSHENZON & al. (1984)
	22d	OH		BEGLEY & al. (1984)

	22e	MOMPON & TOUBIANA (1977)

		R	R′	R″	
	22f	Ac	H	O	JAKUPOVIC & al. (1987 c)
	22g	Ac	H	β-OH, H	JAKUPOVIC & al. (1987 c)
	22h	Meacr	OH	β-OAc, H	JAKUPOVIC & al. (1986 b)

	22i	JAKUPOVIC & al. (1987 c)

		X	
	23a	2α-H	O	GOVINDACHARI & al. (1970)
	23b	2β-H	O	GOVINDACHARI & al. (1972)
	23c	2α-H	α-OMe, H	JAKUPOVIC & al. (1987 b)

	23d	DE SILVA & al. (1982)

		R	X	2-H	
	23e	Meacr	O	α	KUPCHAN & al. (1969)
	23f	iBu	O	α	RUSTAIYAN & al. (1978)
	23g	Sen	O	α	KUPCHAN & al. (1969)
	23h	Meacr	O	β	JAKUPOVIC & al. (1987 b)
	23i	Sen	O	β	JAKUPOVIC & al. (1987 b)
	23j	Tigl	O	β	JAKUPOVIC & al. (1987 b)
	23k	Meacr	α-OMe, H	β	JAKUPOVIC & al. (1987 b)
	23l	Sen	α-OMe, H	β	JAKUPOVIC & al. (1987 b)
	23m	Tigl	α-OMe, H	β	JAKUPOVIC & al. (1987 c)

Table 1 continued

	R	
24a	Meacr	Banerjee & al. (1986)
24b	Tigl	Banerjee & al. (1986)

	R	R'	
24c	Meacr	H	Banerjee & al. (1986), Catalan & al. (1986)
24d	Ang	H	Bohlmann & al. (1981 b)
24e	Ang	OH	Bohlmann & al. (1981 i), Banerjee & al. (1986)

	R	
24f	Ang	Bohlmann & al. (1981 b)
24g	Meacr	Barros & al. (1985)

	R	R'	
24h	Meacr	H	Banerjee & al. (1986), Bohlmann & al. (1982 k)
24i	Ang	H	Banerjee & al. (1986), Bohlmann & al. (1981 b)
24j	Meacr	Me	Banerjee & al. (1986), Bohlmann & al. (1982 k)

	R	R'	
24k	H	H	Begley & al. (1984)
24l	H	OAc	Begley & al. (1984)
24m	O		Begley & al. (1981), Rustaiyan & al. (1980 a)

	R	
24n	Meacr	Begley & al. (1984)
24o	Ang	Begley & al. (1981), Rustaiyan & al. (1980 a)

24p	Gershenzon & al. (1984)

Table 1 continued

	R	R'	
24q	Meacr	Et	MOMPON & al. (1973)
24r	Meacr	Me	BANERJEE & al. (1986)
24s	Tigl	Me	BANERJEE & al. (1986)

	R	X	
25a	Meacr	CH_2	TOUBIANA & GAUDENER (1967)
25b	Hymeacr	CH_2	TOUBIANA (1969)
25c	iBu	CH_2	KUPCHAN & al. (1968)
25d	Meacr	α-Me, H	JAKUPOVIC & al. (1987c)
25e	Hymeacr	α-Me, H	JAKUPOVIC & al. (1987c)

	R	R'	
26a	H	H	
26b	Ac	OAc	BOHLMANN & al. (1980c)

26c			WARNING & al. (1987)

	R	R'	
27a	H	H	BOHLMANN & al. (1982c), HERZ & GOEDKEN (1982)
27b	Meacr	H	HERZ & GOEDKEN (1982), VICHNEWSKI & al. (1976g)
27c	iBu	H	HERZ & GOEDKEN (1982), BOHLMANN & al. (1982c)
27d	Ang	H	HERZ & GOEDKEN (1982), BOHLMANN & al. (1980b)
27e	Tigl	H	HERZ & GOEDKEN (1982), BOHLMANN & al. (1980d)
27f	Espang	H	HERZ & GOEDKEN (1982), ZDERO & al. (1981)
27g	A	H	HERZ & GOEDKEN (1982), BOHLMANN & al. (1982c)
27h	5-Hytigl	H	HERZ & GOEDKEN (1982), BOHLMANN & al. (1982c)
27i	Meacr	OH	HERZ & GOEDKEN (1982), VICHNEWSKI & al. (1976b)
27j	Ang	OH	HERZ & GOEDKEN (1982), LE QUESNE & al. (1982)
27k	Tigl	OH	LE QUESNE (1982)
271	Epmeacr	OH	JAKUPOVIC & al. (1987c)
27m			BANERJEE & al. (1986)

Table 1 continued

	R	R'	
27n	Meacr	H	HERZ & GOEDKEN (1982)
27o	Ang	H	HERZ & GOEDKEN (1982), BANERJEE & al. (1986), BOHLMANN & al. (1982 i)
27p	Tigl	H	HERZ & GOEDKEN (1982), BANERJEE & al. (1986), BOHLMANN & al. (1980 d)
27q	Meacr	OH	HERZ & GOEDKEN (1982), BANERJEE & al. (1986), BOHLMANN & al. (1982 c)
27r	iBu	OH	HERZ & GOEDKEN (1982), BANERJEE & al. (1986), BOHLMANN & al. (1982 c)
27s	Epang	OH	HERZ & GOEDKEN (1982), BANERJEE & al. (1986), BOHLMANN & al. (1982 c)

	R	X	
27t	Meacr	β-OH, H	HERZ & GOEDKEN (1982), BOHLMANN & al. (1981 i)
27u	Ang	β-OH, H	HERZ & GOEDKEN (1982), CATALAN & al. (1986)
27v	Ang	α-OH, H	HERZ & GOEDKEN (1982), BEVELLE & al. (1982), McPHAIL & al. (1975) MANCHAND & al. (1983)

	R	
28a	CH_2 / CH_3	LE QUESNE & al. (1978)
28b	CH_3 / CH_3	RAFFAUF & al. (1975)
28c	CH_3 / H_3C H	MAHMOUD & al. (1980), ZDERO & al. (1981)
28d	H / H_3C CH_3	BOHLMANN & al. (1980 d)
28e	CH_3 / H_3C O H	ZDERO & al. (1981)
28f	H / H_3C O CH_3	BOHLMANN & al. (1982 c)
28g	CH_3 / CH_3	LE QUESNE & al. (1982)

Table 1 continued

		R	R'	R"	
	28 h	CH₂, CH₃ (isopropenyl)	H	OH	BOHLMANN & al. (1982 e)
	28 i	CH₃ / H₃C, H	H	OH	CATALAN & al. (1986)
	28 j	H₃C, CH₃	H	OH	BOHLMANN & al. (1982 c)
	28 k	CH₃ / H₃C, H	H	OAc	CATALAN & al. (1986)
	28 l	CH₃ / H₃C, H	Me	H	MAHMOUD & al. (1980)
	28 m	CH₃ / H₃C, H	Et	H	MAHMOUD & al. (1980)
	28 n	CH₃ / H₃C, H	Et	OH	MAHMOUD & al. (1980)
	28 o	CH₃ / H₃C, H	Me	OH	BARROS & al. (1983 a)
	28 p				BOHLMANN & al. (1983 a)

		R			
	28 q	CH₂, CH₃ (isopropenyl)			BANERJEE & al. (1986)
	28 r	CH₃, CH₃ (isopropyl)			BANERJEE & al. (1986)
	28 s	CH₃ / H₃C, H			BANERJEE & al. (1986)
	28 t	H / H₃C, CH₃			BANERJEE & al. (1986)

		R	R'	
	29 a	H	COOH	BANERJEE & al. (1986)
	29 b	OAc	CH₂OAc	JAKUPOVIC & al. (1987 c)

		R	X	
	29 c	H	β-Me, H	JAKUPOVIC & al. (1987 c)
	29 d	OH	CH₂	JAKUPOVIC & al. (1987 c)

Table 1 continued

		R	X	
	29e	H	O	Kupchan & al. (1968)
	29f	Hymeacr	O	Kupchan & al. (1969)
	29g	Meacr	O	Jakupovic & al. (1986 b)
	29h	Epmeacr	O	Jakupovic & al. (1986 b)
	29i	Hymeacr OHymeacr	O	Jakupovic & al. (1986 b)
	29j	Hymeacr	β-OH, H	Jakupovic & al. (1986 b)

	29k		Ganjian & al. (1983)

		R	
	29l	Hymeacr	Asaka & al. (1977)
	29m	H	Jakupovic & al. (1986 d)

	29n	Lee & al. (1973)

	29o	McPhail & al. (1974)

	29p	Bevelle & al. (1981)

		R	R′	X	
	30a	H	H	H$_2$	
	30b	H	H	α-OH, H	Bohlmann & al. (1978 a)
	30c	H	H	β-OH, H	Bohlmann & al. (1981 e)
	30d	H	H	β-OSen, H	Bohlmann & al. (1981 e)
	30e	H	H	β-OiVal, H	Bohlmann & al. (1979 a)
	30f	H	H	β-OGlu, H	Kupchan & al. (1966)
	30g	H	H	O	
	30h	H	OH	O	Bohlmann & al. (1979)
	30i	OH	H	β-OH, H	
	30j	OH	H	O	Bohlmann & al. (1981 g)
	30k	OMeacr	H	β-OH, H	
	30l	OHymeac	H	β-OH, H	
	30m	OSen	H	β-OH, H	Bohlmann & al. (1982 a)
	30n	04-Hyse	H	β-OH, H	Bohlmann & al. (1982 a)
	30o	OiVal	H	H$_2$	Bohlmann & al. (1980 e)
	30p	OSen	H	H$_2$	Bohlmann & Zdero (1977 a)
	30q	OAc	H	β-OH, H	Jakupovic & al. (1987 c)

Table 1 continued

		R				
	31a	H				CORBELLA & al. (1974), VICHNEWSKI & al. (1980)
	31b	OSen				BOHLMANN & al. (1978a)
	31c	OiVal				BOHLMANN & al. (1980e)

	32	BOHLMANN & al. (1984a)

	33	BOHLMANN & al. (1984g)

	34	BOHLMANN & al. (1981f)

		R		
	35a	H		BOHLMANN & al. (1981e)
	35b	OH		BOHLMANN & al. (1978a)

	36	HERZ & al. (1980), VICHNEWSKI & al. (1977)

	37	ASAKAWA & al. (1981), VICHNEWSKI & al. (1977)

	38	

		R	R′	X	
	39a	OH	H	β-OH, H	BOHLMANN & al. (1981g)
	39b	H	H	O	
	39c	H	OH	O	HOLUB (pers. comm.)
	39d	H	OH	αOH, H	HOLUB (pers. comm.)

Table 1 continued

	40				Bohlmann & al. (1978 a, 1981 d)
	41				Bohlmann & al. (1982 c)
		R			
	42a	Meacr			Bohlmann & al. (1980 e)
	42b	iBu			Bohlmann & al. (1980 d)
	42c	Tigl			Bohlmann & al. (1980 e)
	43				Bohlmann & Zdero (1977 a)
	44				Bohlmann & al. (1981 b)
	45				Bohlmann & al. (1980 e)
	46				Mompon & Toubiana (1976), Narain (1977)

		R	R′	R″	
	47a	B	H	Ac	Bohlmann & Czerson (1978 b), Jakupovic & al. (1987 a)
	47b	5-Acang	OH	Ac	Bohlmann & Czerson (1978 b), Jakupovic & al. (1987 a)
	47c	B	H	H	Jakupovic & al. (1987 a)
	47d	4,5-Diacang	H	Ac	Jakupovic & al. (1987 a)
	47e	C	H	Ac	Jakupovic & al. (1987 a)
	47f	4-Acang	H	H	Jakupovic & al. (1987 a)
	47g	5-Acang	H	Ac	Jakupovic & al. (1987 a)
	47h	4-Acang	H	Ac	Jakupovic & al. (1987 a)
	47i	4,5-Diacang	OH	Ac	Jakupovic & al. (1987 a)
	47j	5-Acang	Olt	Ac	Jakupovic & al. (1987 a)

		X	
	48a	CH$_2$	Jakupovic & al. (1987 a)
	48b	β-Me, H	Jakupovic & al. (1987 a)

Table 1 continued

		R		
	49a	H		BOHLMANN & al. (1980 e)
	49b	OiVal		BOHLMANN & al. (1980 e)

50

		X		
	51a	O		JAKUPOVIC & al. (1987 c)
	51b	β-OH, H		JAKUPOVIC & al. (1987 c)

		R		
	52a	H		HERZ & al. (1981)
	52b	Et		HERZ & al. (1981)

53   NARAIN (1978)

54   NARAIN (1977)

5-Alkylcoumarins

55

	56a	56b	56c
R	H	OMe	H
R'	H	H	OMe
R"	Me	H	H

	57a	57b	57c	57d	57e
R	H	H	Me	Me	Me
R'	Ac	Prop	H	Ac	Prop

	58a	58b
R	H	Me

55	MAHMOUD & al. (1980)
56a	BOHLMANN & ZDERO (1977 d)
56b	BOHLMANN & ZDERO (1977 d)
56c	BOHLMANN & ZDERO (1977 d)
57a	JAKUPOVIC & al. (1986 d)
57b	JAKUPOVIC & al. (1986 d)
57c	JAKUPOVIC & al. (1986 d)
57d	JAKUPOVIC & al. (1986 d)
57e	JAKUPOVIC & al. (1986 d)
58a	JAKUPOVIC & al. (1986 d)
58b	JAKUPOVIC & al. (1986 d)

F. BOHLMANN & J. JAKUPOVIC:

Table 1 continued

59a 59b

R   Ac Prop

60a 60b 60c

R   H   OMe H
R′   H   H   OH

59a JAKUPOVIC & al. (1986 d)
59b JAKUPOVIC & al. (1986 d)
60a BOHLMANN & ZDERO (1977 c)
60b BOHLMANN & ZDERO (1977 c)
60c BOHLMANN & ZDERO (1977 c)

61a 61b 61c 61d   62   63

R   H   Me   H   H
R′   H   H   OMe OH

61a BOHLMANN & ZDERO (1977 d)
61b BOHLMANN & ZDERO (1977 d)
61c BOHLMANN & ZDERO (1977 d)
61d BOHLMANN & ZDERO (1977 d)
62   RUSTAIYAN & al. (1980 b)
63   RUSTAIYAN & al. (1980 b)
64   BOHLMANN & ZDERO (1977 c)
65a BOHLMANN & ZDERO (1977 d)
65b BOHLMANN & ZDERO (1977 d)
66   BOHLMANN & ZDERO (1977 f)
67a BOHLMANN & ZDERO
     (1982 f, 1977 b),
     JAKUPOVIC & al. (1987 a)
67b JAKUPOVIC & al. (1987 a)
68   BALBAA & al. (1980 a),
     BOHLMANN & al. (1981 g)
69   BALBAA & al. (1980 b)

64   65a 65b   66

                 R   H   Me

67a   R = H   68   69
67b   R = Me

Table 1 continued

70   71a   71b (E)   72

70	SHUKLA & al. (1982)
71a	SHUKLA & al. (1982)
71b	SHUKLA & al. (1982)
72	BOHLMANN & ZDERO (1977b)
73	BALBAA & al. (1980b)
74	BALBAA & al. (1980a)
75	JAKUPOVIC & al. (1987a)
76a	BALBAA & al. (1980b)
76b	JAKUPOVIC & al. (1987a)
76c	JAKUPOVIC & al. (1987a)

73   74

75   76a

76b   R = H
76c   R = Me

Sesquiterpenes

Nerolidol derivatives

77
78
79
80
81
82
83
84
85

77	BOHLMANN & al. (unpubl.)
78	BOHLMANN & al. (unpubl.)
79	BOHLMANN & al. (unpubl.)
80	BOHLMANN & al. (1980b)
81	BOHLMANN & al. (1981c)
82	BOHLMANN & al. (1981c)
83	BOHLMANN & al. (1981c)
84	BOHLMANN & al. (unpubl.)
85	BOHLMANN & al. (1983a)

Table 1 continued

Bisabolane derivatives

	86a	86b	86c	86d	86e
R	H	H	OH	H	H
R'	H	H	H	OH	OH
X	H_2	O	O	H_2	H_2
					10, 11 H

87   88a   88b   89

	88a	88b
R	OMe	H
R'	H	OSen

Further sesquiterpenes

90a 90b 90c   91a   91b   91c   92a 92b 92c

| R | H | OH | H,$^{\Delta 3}$ | R | CO_2H | CHO | CO_2H,4epi | 3 | 4 | 4(15) |

93a   93b   94   95a   R = CHO,   R' = Me
     (11, 13 H)   95b   R = Me,   R' = CHO

96a   96b   96c   96d   97

| R | H | OH | H(5,6 H) | OAc |

98a   98b
    2-epi

86a	Bohlmann & al. (1982 k)
86b	Bohlmann & al. (1982 k)
86c	Bohlmann & al. (1982 k)
86d	Bohlmann & al. (1983 a)
86e	Bohlmann & al. (1983 a)
87	Bohlmann & al. (1981 e)
88a	Bohlmann & al. (1986 d)
88b	Bohlmann & al. (1986 d)
89	Bohlmann & al. (1986 d)

90a	Bohlmann & al. (1982 k)
90b	Bohlmann & al. (1982 k)
90c	Bohlmann & al. (1982 k)
91a	Bohlmann & al. (1982 b)
91b	Bohlmann & al. (1983 a)
91c	Bohlmann & al. (1982 b)
92a	Bohlmann & al. (1983 a)
92b	Bohlmann & al. (1983 a)
92c	Bohlmann & al. (1983 a)
93a	Bohlmann & al. (1981 c)
93b	Bohlmann & al. (1981 c)
94	Bohlmann & al. (1984 b)
95a	Bohlmann & Zdero (1982 f)
95b	Bohlmann & Zdero (1982 f)
96a	Bohlmann & al. (1980 b), Vichnewski & al. (1980)
96b	Bohlmann & al. (1980 b)
96c	Bohlmann & al. (1980 b)
96d	Bohlmann & al. (1980 b)
97	Bohlmann & al. (1981 d)
98a	Bohlmann & al. (1983 e)
98b	Bohlmann & al. (1983 c)

Table 1 continued

99a   Me
99b   CH$_2$O
99c   CH$_2$OAc
99d   CHO
99e   CO$_2$H

100a   100b   100c

R   Me   CH$_2$OH   CHO

101

102

103a   103b
     (10 epi)

104

105

106

107

99a	BOHLMANN & al. (1982 i)
99b	BOHLMANN & al. (1982 i)
99c	BOHLMANN & al. (1982 i)
99d	BOHLMANN & al. (1982 i)
99e	BOHLMANN & al. (1982 i)
100a	BOHLMANN & al. (1982 i)
100b	BOHLMANN & al. (1982 i)
100c	BOHLMANN & al. (1982 i)
101	BOHLMANN & al. (1982 i)
102	BOHLMANN & al. (1982 i)
103a	JAKUPOVIC & al. (1986 d)
103b	BOHLMANN & al. (1984 b)
104	BOHLMANN & al. (1984 b)
105	BOHLMANN & al. (1984 b)
106	BOHLMANN & al. (1986 d)
107	BOHLMANN & al. (1984 b)

Corymbium − Diterpenes

108	ZDERO & BOHLMANN (1988)
109	ZDERO & BOHLMANN (1988)
110	ZDERO & BOHLMANN (1988)
111	ZDERO & BOHLMANN (1988)
112	ZDERO & BOHLMANN (1988)

108   R = H, Ac, COCH$_2$CO$_2$H   109   110   ($\Delta^{8(17)}$, 6 β OH)
       R′ = H, Ac, iVal, Meval

111   R = H   112   R = Ac

Table 2. Abbreviation of common esters

Abbrev	Structure	Abbrev	Structure
Ac		5-Hysen	
4-AcAng		4-Hytigl	
5-AcAng		5-Hytigl	
4.5 Diacang		Meacr	
5-Acsen		Pro	
iBu		Sen	
Capr		Tigl	
Epang		iVal	
Epmeacr		A	
Oglu	O glucopyranoside	B	
Hymeacr		C	
4-Hysen		D	

Table 3. Distribution of characteristic compounds in the tribe *Vernonieae* (ROBINSON & al. 1980)

Taxon	Compound	References
Pseudostifftinae		
Pseudostifftia kingii H. ROBINS.	30 i−m, 35 b, 80	BOHLMANN & al. (1982 a)
Stokesia laevis (HILL) GREENE	9 h, 9 p, 9 r, 30 e, 30 h	BOHLMANN & al. (1979)
Lychnophorinae		
Albertinia brasiliensis SPRENG.	−	BOHLMANN & al. (1980 e)
Chresta sphaerocephala DC.	4 c, 9 o, 27 e	BOHLMANN & al. (1981 f, 1982 c)
Eremanthus bicolor (SCH. BIP.) BAKER	21 a, 27 b, 27 e, 27 n, 27 p, 28 a/b, 28 d, 28 q, 28 r, 28 t, 30 a−g, 31 a, 38	BOHLMANN & al. (1980 d)
E. crotonoides (DC.) SCH. BIP.	27 b−e, 27 n, 27 r, 28 a, 28 b, 28 d	BOHLMANN & al. (1982 c)
E. elaeagus SCH. BIP.	28 a/b, 28 g, 31 a	LE QUESNE & al. (1978), RAFFAUF & al. (1975), VICHNEWSKI & al. (1972)
E. glomerulatus LESS.	21 a, 24 d, 24 f, 24 i, 27 a, 27 f−h, 27 s, 28 c−f, 28 j	BOHLMANN & al. (1981 b, 1982 c)
E. goyazensis SCH. BIP.	27 i, 36, 37	VICHNEWSKI & al. (1976 b, 1977)
E. incanus LESS.	21 a, 28 b−g, 31 a, 36, 37	BOHLMANN & al. (1980 d), HERZ & al. (1980)
E. mollis SCH. BIP.	27 i, 27 j	BOHLMANN & al. (1981 b)
Lychnophora affinis GARDN.	27 j, 27 k, 28 a, 28 c, 28 l−n, 96 a/b	MAHMOUD & al. (1980)
L. bahiensis MATTF.	27 b, 27 d, 27 j, 28 c, 28 i, 28 k, 30 a, 31 a	BOHLMANN & al. (1982 k)
L. blanchetii (SCH. BIP.) H. ROBINS.	21 a, 22 a, 24 c, 24 h−j, 27 a/b, 27 d, 31 a, 42 a, 42 c, 45	BOHLMANN & al. (1980 e, 1982 k)
L. columnaris MATTF.	27 b, 31 a, 34, 39 b, 99 a−e, 100 a−e, 101, 102	BOHLMANN & al. (1982 i)
L. crispa MATTF.	27 d/e, 27 o, 27 p, 28 c, 28 i, 28 s/t, 31 a	BOHLMANN & al. (1982 k)
L. ericoides GARDN.	21 a, 27 d, 31 a	BOHLMANN & al. (1980 b)
L. martiana GARDN.	96 a/b	VICHNEWSKI & al. (1980)
L. passerina GARDN.	21 a, 27 i, 30 a, 31 a, 34, 38, 39 b, 106	BOHLMANN & al. (1981 d)
L. salicifolia MART.	21 a, 31 a, 96 a−d, 97, 98 a/b	BOHLMANN & al. (1980 b, 1981 d, 1983 c)
L. selowii SCH. BIP.	27 d, 27 j, 27 u, 27 v, 31 a, 86 a−c	BOHLMANN & al. (1982 k)
L. uniflora SCH. BIP.	27 d, 28 a, 28 c, 31 a	BOHLMANN & al. (1981 d),
Piptolepis ericoides (LESS.) SCH. BIP.	24 e, 27 o	BOHLMANN & al. (1981 i)
P. leptospermoides (DC.) SCH. BIP.	24 i, 27 b, 27 d, 28 a, 28 c, 28 h, 31 a	BOHLMANN & al. (1982 e)
Proteopsis argentea MART. ex ZUCC.	27 d, 27 f, 28 c, 28 e	ZDERO & al. (1981)

Table 3 continued

Taxon	Compound	References
Vanillosmopsis brasiliensis (Gardn.) Sch. Bip.	27 i, 27 t, 31 a	Bohlmann & al. (1981 i)
V. discolor Benth. & Hook.	–	Bohlmann & al. (1981 i)
V. erythropappa Sch. Bip.	27 d	Vichnewski & al. (1976)
V. pohlii Baker	27 i, 27 t	Bohlmann & al. (1981 i)
Centratherinae		
Centratherum punctatum Cass.	21 g, 27 i/j, 27 l, 27 v, 28 b, 29 a, 30 o, 31 a, 31 c	Banerjee & al. (1986), Bevelle & al. (1981), Bohlmann & al. (1980 e), Jakupovic & al. (1987 c), Manchand & al. (1983), Ohno & al. (1979)
Piptocarphinae		
Piptocarpha chontalensis Pall.	9 b, 9 g, 9 j, 9 k, 9 m, 9 t	Cowall (1981)
P. oblanga (Gardn.) Baker	–	Bohlmann & al. (1980 e)
P. opaca Baker	9 u, 9 v	Herz & Kulanthaivel (1983)
P. poeppigiana Baker	9 j	Bohlmann & al. (unpubl.)
P. rotundifolia (Less.) Baker	–	Bohlmann & al. (unpubl.)
Critoniopsis bogotana (Cuatr.) H. Robins.	3 c, 3 e, 3 h, 5 f, 5 h, 5 i, 8 g, 8 j, 8 k	Jakupovic & al. (1986 a)
Vernoniinae		
Bothriocline laxa N. E. Br.	56 a, 56 c, 60 a/d, 61 a – d, 65 a/b	Bohlmann & Zdero (1977 d)
B. longipes (Oliv. & Hiern.) N. E. Br.	57 a – e, 58 a/b, 59 a/b	Jakupovic & al. (1987 a)
Ethulia conyzoides L.	55, 67 a, 68, 69, 70, 71 a/b, 72, 73, 74, 76 a	Balbaa & al. (1980 a, b), Bohlmann & al. (1977 c, 1981 a), Mahmoud & al. (1980), Shukla & al. (1982)
Volkensia ripensis Hutch.	67 a/b, 75	Jakupovic & al. (1987 a)
Erlangea cordifolia S. Moore	22 d, 24 k – o	Begley & al. (1981, 1984)
E. fusca S. Moore	62, 63	Rustaiyan & al. (1980 b)
E. inyangana (N. E. Br.) B. L. Burtt	47 a/b	Bohlmann & Czerson (1978 b)
E. renifolia Wild & Pope	50	Bohlmann & Czerson (1978 c)
E. rogersii S. Moore	60 a – c, 64 a/b	Bohlmann & Zdero (1977 b)
Gutenbergia marginata (Oliv. & Hiern.) Wild & Pope	47 a – j, 48 a/b	Jakupovic & al. (1987 a)
Heterocoma albida DC.	31 a	Herz & al. (1980)
Hoplophyllum spinosum DC.	30 c	Bohlmann & al. (unpubl.)
Mattfeldanthus nobilis (H. Robins.) H. Robins.	26 b	Bohlmann & al. (1980 c)
Gongrothamnus auriantiaca N. E. Br.	29 c/d	Jakupovic & al. (1987 c)
G. sublutea S. Elliot	46	Mompon & Toubiana (1977)
Struchium sparganophora (L.) O. Kuntze	51	Jakupovic & al. (1987 c)

Table 3 continued

Taxon	Compound	References
Stilpnopappus glomeratus GARDN.	4 a – c	BOHLMANN & al. (1982 h)
S. pickelii MATTF.	5 f, 5 i	BOHLMANN & al. (1982 h)
S. tomentosus MART. in DC.	5 a/b, 5 d – f, 5 i, 8 g, 8 k	BOHLMANN & al. (1982 h)
S. trichospiroides MART.	5 d	BOHLMANN & al. (1982 h)
Muschleria stolzii S. MOORE	2 a, 2 i	JAKUPOVIC & al. (1987 c)
Alcantara ekmaniana (PHILIPSON) H. ROBINS.	27 b, 27 d, 27 f, 27 o, 90 a, 90 c, 91 a, 91 c	BOHLMANN & al. (1982 b)
Oliganthes discolor SCH. BIP.	27 i – k	BOHLMANN & al. (1982 b)
Corymbium villosum L.	108, 109, 110, 111, 112	ZDERO & BOHLMANN (1988)

Elephantopinae

Taxon	Compound	References
Elephantopus angustifolius Sw.	22 b, 23 e, 23 h – m	JAKUPOVIC & al. (1987 b)
E. carolinianus WILLD.	32, 33	BOHLMANN & al. (1984 g)
E. elatus BERTOL.	23 e, 23 g	KUPCHAN & al. (1966)
E. hirtiflorus DC.	49 a/b	BOHLMANN & al. (1980 e)
E. mollis HBK	23 a – c, 24 a/b, 24 r/s	BANERJEE & al. (1986), BOHLMANN & ZDERO (1976 a), JAKUPOVIC & al. (1987 b)
E. scaber L.	23 d	DE SILVA & al. (1982), GOVINDACHARI & al. (1972)
E. tomentosus L.	23 f	RUSTAIYAN & al. (1978)
Pseudoelephantopus spicatus (B. JUSS.) ROHR	2 a, 51, 18 a, 50, 89	JAKUPOVIC & al. (1986 d)
P. spiralis (LESS.) CRONQ.	30 d	BOHLMANN & al. (unpubl.)

Rolandrinae

Taxon	Compound	References
Rolandra fruticosa (L.) KUNTZE	7 a/b, 52 a/b	HERZ & al. (1981)

Vernonia

I. Subg. *Orbivestus* S. B. JONES

I. 1. Sect. *Orbivestus*

a. Subsect. *Orbivestus*

Taxon	Compound	References
V. cinarescens SCH. BIP.	66, 67 a	BOHLMANN & ZDERO (1982 f)
V. galpinii KLATT	60 a, 64, 86 c	BOHLMANN & ZDERO (1982 f)
V. cistifolia O. HOFFM.	6 a, 6 d, 6 g, 6 m, 6 n	JAKUPOVIC & al. (1987 c)
V. colorata DRAKE	25 a/b	TOUBIANA (1969), TOUBIANA & GAUDENER (1967)
V. holstii O. HOFFM.	6 q	JAKUPOVIC & al. (1987 c)
V. monocephala HARV.	7 d, 8 a, 8 c	BOHLMANN & al. (1982 g)
V. pectoralis BAKER	21 e/f, 22 e, 25 a, 29 a	McPHAIL & al. (1975), MOMPON & al. (1974), MOMPON & TOUBIANA (1976)

Table 3 continued

Taxon	Compound	References
b. Subsect. *Strobacalyx* S. B. Jones		
V. amygdalina Delile	25 a, 25 c, 29 f, 29 k/l	Ganjian & al. (1983), Kupchan & al. (1969)
V. arborea Buch.-Ham.	–	Bohlmann & Zdero (1977 a)
V. brachycalyx O. Hoffm.	6 r/s	Jakupovic & al. (1987 c)
V. conferta Benth.	5 c, 5 j, 5 l, 5 m	Bohlmann & al. (unpubl.), Toubiana & al. (1972)
V. frondosa Oliv. & Hiern	5 l, 5 n	Jakupovic & al. (1987 c)
V. mespilifolia Less.	29 p	Bohlmann & al. (unpubl.)
c. Subsect. *Pawelianae* S. B. Jones		
V. anisochaetoides Sond.	30 b, 30 l	Bohlmann & al. (1978 a)
V. angulifolia DC.	8 e–h	Bohlmann & al. (1978 a)
V. tufnellae S. Moore	25 a/b	Jakupovic & al. (1987 c)
d. Subsect. *Hillardianae* S. B. Jones		
V. dregeana Sch. Bip. ex Walp.	–	Bohlmann & al. (1978 a)
V. hirsuta Sch. Bip. ex Walp.		
var. *flanganii*	21 a–c, 30 a, 30 d, 86 c	Bohlmann & al. (1978 a)
var. *hirsuta*	8 g–i, 21 a–c, 30 a, 30 d, 31 a	Bohlmann & al. (1978 a)
V. natalensis Sch. Bip. ex Walp.	1 c/d, 2 e, 3 c, 4 o	Bohlmann & al. (1978 a), Bohlmann & Zdero (1982 f)
V. oligocephala (DC.) Sch. Bip. ex Walp.	3 d, 5 g, 8 g, 8 m, 30 p, 31 b	Bohlmann & al. (1978 a, 1984 b)
V. smithiana Less.	5 o, 5 y	Jakupovic & al. (1987 c)
V. capensis (Houtt.) Druce	5 o	Bohlmann & al. (unpubl.)
V. poskeana Vatke & Hild.	5 u/v, 7 i, 10 h/i	Jakupovic & al. (1986 a)
var. *chlorolepsis*	10 g/h	Bohlmann & al. (1983 a)
var. *poskeana*	10 c–f, 28 p	Bohlmann & al. (1983 a)
V. stipulacea Klatt	95 a/b	Bohlmann & Zdero (1982 f)
e. Subsect. *Centrapalus* S. B. Jones		
V. fastigiata Oliv. & Hiern	79	Bohlmann & al. (unpubl.)
V. praemorsa Muschler	25 a, 29 f, 29 j	Jakupovic & al. (1987 c)
V. sutherlandii Harv.	2 h, 3 c/d, 104, 105	Bohlmann & al. (1984 b)
I. 2. Sect. *Stengelia*		
a. Subsect. *Stengelia*		
V. adoensis Sch. Bip. ex Walp.	6 o/p, 94	Bohlmann & al. (1984 b)
V. anthelmintica (L.) Willd.	29 l	Asaka & al. (1977)
V. hymenolepis A. Rich.	29 e/f, 29 k, 29 n	Lee & al. (1973), Bohlmann & al. (unpubl.)
V. lasiopus O. Hoffm.	29 e–g, 29 l/m	Jakupovic & al. (1987 c)

Table 3 continued

Taxon	Compound	References
I. 3. Sect. *Tephrodes*		
a. Subsect. *Tephrodes*		
V. chinensis (LAM.) LESS.	21 a, 30 a, 30 d, 30 g, 31 a	BOHLMANN & al. (1981 d)
V. chloropappa BAKER	4 d − g, 5 s	JAKUPOVIC & al. (1987 c)
V. cinerea LESS.	2 f, 2 i, 5 i, 5 y, 8 i, 8 l, 9 w, 9 x	JAKUPOVIC & al. (1986 a)
I. 4. Sect. *Azureae*		
V. glabra VATKE	21 h, 22 h, 29 e − j, 29 l − n, 30 a, 31 a, 31 b, 92 a − c	BOHLMANN & al. (1983 a), JAKUPOVIC & al. (1986 b, 1987 c)
V. staehelinoides HARV.	5 u	BOHLMANN & al. (1982 d)
V. melleri OLIV. & HIERN	29 f	JAKUPOVIC & al. (1987 c)
II. Subg. *Vernonia*		
II. 1. Sect. *Vernonia*		
a. Subsect. *Novebaracensis*		
V. acaulis (WALT.) GLEASON	5 l, 5 n	ABDUL-BASET & al. (1971), MABRY & al. (1975)
V. alamanii DC.	5 n	ABDUL-BASET & al. (1971), MABRY & al. (1975)
V. angustifolia MICHX.	5 l, 5 n	MABRY & al. (1975)
V. arkansana DC.	2 b − d, 14 a − c, 16, 17 a/b, 21 a, 30 a, 30 c/d, 30 g, 30 j, 31 a, 35 b	BOHLMANN & al. (1981 c), BOHLMANN & al. (1981 g), PADOLINA & al. (1974 a)
V. baldwinii TORR.	5 l, 5 n	MABRY & al. (1975), PADOLINA & al. (1974 b)
V. blodgettii SMALL	5 n	MABRY & al. (1975), ABDUL-BASET & al. (1971)
V. cronquistii S. B. JONES	−	BOHLMANN & al. (unpubl.)
V. fasciculata MICHX.	2 d, 5 l, 5 n, 53, 54	BOHLMANN & al. (unpubl.), NARAIN (1977), NARAIN (1978), PADOLINA & al. (1974 g)
V. flaccidifolia SMALL	2 d, 5 l, 5 n	ABDUL-BASET & al. (1971)
V. gigantea (WALT.) TREL.	15 n, 5 l	ABDUL-BASET & al. (1971), BOHLMANN & al. (1981 e)
V. glauca (L.) WILLD. var. glabra	5 l, 5 n	ABDUL-BASET & al. (1971), PADOLINA & al. (1974 b)
V. greggii GRAY	5 n, 5 q, 9 b, 9 j/k	DOMINGUEZ & al. (1986), JAKUPOVIC & al. (1974 b)
V. larsenii KING & JONES	−	BOHLMANN & al. (unpubl.)
V. lettermannii ENGELM.	5 n, 5 p/q	ABDUL-BASET & al. (1971), JAKUPOVIC & al. (1987 c)
V. liatroides DC.	2 f/g, 5 n	BOHLMANN & al. (unpubl.), MALDONADO & al. (1980)
V. lindheimeri GRAY & ENGELM.	5 l, 5 n	ABDUL-BASET & al. (1971)

Table 3 continued

Taxon	Compound	References
V. marginata (Torr.) Raf.	2 d, 2 k, 3 h/i, 5 t, 5 w, 5 x, 19 a−d, 20	Jakupovic & al. (1986 c), Padolina & al. (1974 a)
var. *marginata*	2 d, 5 l, 5 n	Abdul-Baset & al. (1971)
var. *tenuifolia*	2 d, 5 l, 5 n	Abdul-Baset & al. (1971), Padolina & al. (1974 a)
V. missurica Raf.	5 n	Mabry & al. (1975)
V. novebaracensis (L.) Michx.	2 d, 5 l, 5 n, 10 b, 30 c, 30 g, 35 b	Abdul-Baset & al. (1971)
V. oaxacana Sch. Bip. ex Klatt	−	Bohlmann & al. (unpubl.)
V. pulchella Small	−	Bohlmann & al. (unpubl.)
V. serratuloides HBK	51	Bohlmann & al. (unpubl.)
V. texana (Gray) Small	5 l, 5 n	Abdul-Baset & al. (1971)
b. Subsect. *Polyanthes*		
V. deppeana Less.	5 n	Bohlmann & al. (unpubl.)
V. menthaefolia Less.	hirsutinolides	Holub (pers. comm.)
V. patens HBK	3 c, 3 e, 5 h, 7 e−g, 8 b, 8 j, 12 a, 12 c, 12 e	Jakupovic & al. (1986 d)
V. polyanthes (Spreng.) Less.	9 s, 10 b, 81, 82, 83	Bohlmann & al. (1981 c)
c. Subsect. *Eremosis*		
V. bogotana Cuatr. (= *Critoniopsis*)		
V. duncanii S. B. Jones	51	Mabry & al. (1975)
V. leiocarpa DC.	2 c	Betkouski & al. (1976)
V. obtusa Blake	−	Mabry & al. (1975)
V. salicifolia (DC.) Sch. Bip.	2 f/g	Mabry & al. (1975)
V. steetzii Sch. Bip.	−	Mabry & al. (1975)
V. triflosculosa HBK	43, 50	Bohlmann & Zdero (1977 a)
V. uniflora Sch. Bip.	2 c, 2 f/g, 5 n	Betkouski & al. (1975), Mabry & al. (1975)
d. Subsect. *Chamaedrys*		
V. chamaedrys Less.	30 d	Bohlmann & al. (unpubl.)
V. jalcana Cuatr.	2 a, 3 a−c, 5 l, 5 n, 9 a, 9 c−h, 12 b, 12 d, 12 h, 13 a−d	Jakupovic & al. (1986 d)
V. laxa Gardn.	−	Bohlmann & al. (1981 e)
V. myrsitinitis Ekman	−	Bohlmann & al. (1981 e)
V. tomentella Mart.	−	Bohlmann & al. (1981 e)
e. Subsect. *Stenocephalum*		
V. megapotamica Spreng.	−	Bohlmann & Zdero (1977 a)
f. Subsect. *Nudiflorae*		
V. accunae Alain	5 n, 9 j	Holub (pers. comm.)
V. acuminata Less.	51	Jakupovic & al. (1986 d)

Table 3 continued

Taxon	Compound	References
V. angusticeps EKM.	reynosin, 39 c/d	HOLUB (pers. comm.)
V. arborescens (L.) Sw.	5 n	MABRY & al. (1975)
V. aurea MART. ex DC.	5 n, 9 p, 9 r, 21 a, 88 a	JAKUPOVIC & al. (1986 d)
V. bardanoides LESS.	5 l	BOHLMANN & al. (unpubl.)
V. brachiata BENTH.	5 l	JAKUPOVIC & al. (1986 d)
V. brevifolia LESS.	5 l	MABRY & al. (1975)
V. buddleiaefolia MART. ex DC.	9 r	BOHLMANN & al. (unpubl.)
V. canescens HBK	5 l	MABRY & al. (1975)
V. compactiflora MART. ex BAKER	8 a, 8 c	BOHLMANN & al. (1982 g)
V. cognata LESS.	9 h	BOHLMANN & al. (unpubl.)
V. cotoneaster LESS.	1 a/b, 11 a−e, 27 b	BOHLMANN & al. (1980 a, 1982 g, JAKUPOVIC & al. (1986 d)
V. coriacea LESS.	−	BOHLMANN & al. (1982 h)
V. echitifolia MART. ex DC.	30 g, 35 a, 87	BOHLMANN & al. (1981 e)
V. elegans GARDN.	−	BOHLMANN & al. (1981 e)
V. farinosa BAKER in MART.	−	BOHLMANN & al. (1981 e)
V. flexuosa SIMS	30 d, 30 f	KUPCHAN & al. (1966)
V. fruticosa (L.) Sw.	5 l	MABRY & al. (1975)
V. fruticulosa MART. ex DC.	5 l, 9 r	BOHLMANN & al. (unpubl.)
V. holosericea MART. ex DC.	−	BOHLMANN & al. (1981 c)
V. incana LESS.	5 n	MABRY & al. (1975)
V. intermedia DC.	21 a, 30 a, 30 c/d, 30 g, 31 a	BOHLMANN & al. (1981 e)
V. lilacina MART. ex DC.	21, 11 c−e	BOHLMANN & al. (1981 c)
V. mariana MART.	3 c, 7 e, 41, 83, 84	JAKUPUVIC & al. (1986 d)
V. missionis GARDN.	−	BOHLMANN & al. (1987 e)
V. moaensis ALAIN	8 n, 39 d	HOLUB (pers. comm.)
V. nudiflora LESS.	5 n, 21 d, 30 f	BOHLMANN & ZDERO (1977 a), MABRY & al. (1975)
V. obtusata LESS.	26 a	BOHLMANN & al. (1981 e)
V. platensis LESS.	−	BOHLMANN & al. (1981 e)
V. pluvialis GLEASON	5 l	BOHLMANN & al. (1981 e)
V. saltensis HIERON.	9 h, 9 p	BOHLMANN & al. (1979 a)
V. scorpioides (LAM.) PERS.	5 n, 5 p, 5 r, 9 h/i, 9 n, 9 p, 11 d, 15, 26 c	BOHLMANN & al. (1979 a), JAKUPOVIC & al. (1986 b), WARNING & al. (1987)
V. tortuosa (L.) BLAKE	5 l	JAKUPOVIC & al. (1987 c)
V. trinitatis EKMAN	5 l	JAKUPOVIC & al. (1987 c)
V. tweediana BAKER ex MART.	9 b, 9 j	BOHLMANN & al. (unpubl.)
V. venosissima SCH. BIP. ex BAKER	9 l, 10 a	BOHLMANN & al. (1981 e)
V. warmingiana BAKER	−	BOHLMANN & al. (unpubl.)
V. westiniana LESS.	5 n, 5 q	JAKUPOVIC & al. (1987 c)

II. 2. Sect. *Lepidonia*

V. jonesii B. L. TURNER	22 c, 24 p	GERSHENZON & al. (1984)
V. pooleae B. L. TURNER	5 l	GERSHENZON & al. (1984)

Table 3 continued

Taxon	Compound	References
III. Further *Vernonia* spp.		
V. bellinghamii S. MOORE	3 c/d, 3 f, 7 e, 8 d	JAKUPOVIC & al. (1987 c)
V. chalybaea MART.	51	BOHLMANN & al. (1982 g)
V. condensata BAKER	25 a/b, 25 d/e, 29 f, 29 k,	JAKUPOVIC & al. (1987 c)
V. diffusa LESS.	22 f/g, 29 b, 30 q, 51 a/b	JAKUPOVIC & al. (1987 c)
V. fagifolia GARDN.	83	BOHLMANN & al. (1981 c)
V. flexipappa GLEASON	51	JAKUPOVIC & al. (1986 a)
V. fulta SCHREB.	3 a, 5 j−1	CATALAN & al. (1985)
V. guineensis BENTH.	29 e/f	TOUBIANA & al. (1975)
V. kuntzei HIERON.	9 h, 9 r	BOHLMANN & al. (1981 e)
V. lanuginosa GARDN.	93 a/b	BOHLMANN & al. (1981 c)
V. mollissima DON	9 g/h, 9 y	CATALAN & al. (1986)
V. cf. *profuga* L.	2 d, 14 c, 21 a, 30 a, 30 d, 30 g, 30 j, 31 a, 35 b, 39 a	BOHLMANN & al. (1981 g)
V. spec. *n.*	6 b−f, 6 h−l, 7 c	JAKUPOVIC & al. (1987 c)
V. squamulosa H. & ARN.	5 n, 5 p, 9 j	CATALAN & al. (1986)

Table 4. Separation of the tribe *Vernonieae* into subtribes

Tribe *Vernonieae*

Subtribe

Pseudostifftinae	guaianolides
Lychnophorinae	furoheliangolides
(*Chresta*)	glaucolides
Centratherinae	furoheliangolides, guaianolides
Piptocarphinae	glaucolides
Elephantopinae	germacranolides (dilactones)
Rolandrinae	glaucolides
Vernoniinae	glaucolides (in part)

excluded: *Corymbium* (diterpenes)

Table 5. Genera of the subtribe *Vernoniinae* with main constituents

Bothriocline *Ethulia* *Volkensia* *Erlangea* p.p.	5-alkyl-4-hydroxy coumarin derivatives
Erlangea p.p. *Gutenbergia* *Hoplophyllum*	guaianolides
Mattfeldanthus	germacranolides
Gongrothamnus	elemanolides
Vernonia *Struchium* *Stilpnopappus* *Muschleria*	glaucolides, vernolepins
Alcantara	lactone precursors

References

ABDUL BASET, Z., SOUTHWICK, L., PADOLINA, W., YOSHIOKA, H., MABRY, T. J., JONES, S. B., 1971: Sesquiterpene lactones: a survey of 21 United States taxa from the genus *Vernonia* (*Compositae*). − Phytochemistry **10**: 2201−2204.

ASAKA, Y., KUBOTA, T., KULKARNI, A. B., 1977: Studies on a bitter principle from *Vernonia anthelmintica*. − Phytochemistry **16**: 1838−1839.

ASAKAWA, Y., TAIRA, Z., TOYOTA, M., TAKEMOTO, T., HERZ, W., SAKAI, T., 1981: Sesquiterpene lactones of *Eremanthus incanus* and *Porella japonica*. − J. Org. Chem. **46**: 4602−4604.

BALBAA, S. I., BOHLMANN, F., HALIM, A. F., HALAWEISH, F. T., 1980: New coumarin derivatives from *Ethulia conyzoides*. − Planta Med. **39**: 218.

− HALIM, A. F., HALAWEISH, F. T., BOHLMANN, F., 1984: New 5-methylcoumarin derivatives from *Ethulia conyzoides*. − Phytochemistry **19**: 1519−1522.

BANERJEE, S., SCHMEDA-HIRSCHMANN, G., CASTRO, V., SCHUSTER, A., JAKUPOVIC, J., BOHLMANN, F., 1986: Further sesquiterpene lactones from *Elephantopus mollis* and *Centratherum punctatum*. − Planta Med. **52**: 29−32.

BARROS, D. A. D., LOPES, J. L C., VICHNEWSKI, W., LOPES, J. N. C., KULANTHAIVEL, P., HERZ, W., 1985: Sesquiterpene lactones in the molluscidal extract of *Eremanthus glomeratus*. − Planta Med. **51**: 38−39.

BEGLEY, M. J., CROMBIE, L., CROMBIE, W. M. L., GATUMA, A. K., MARADUFU, A., 1981: Germacranolides of *Erlangea cordifolia*: structure and absolute stereochemistry of cordifene and cordifene 4 β,15-oxide by x-ray and spectroscopic methods. − J. Chem. Soc. Perkin Trans. **1**: 2702−2709.

− − − − − 1984: Germacranolides of *Erlangea cordifolia*: isolation and structures of cordifolia-54, 55-P 2, and -31 by spectral and X-ray methods. − J. Chem. Soc. Perkin Trans. **1**: 819−824.

Betkouski, M., Mabry, T. J., Taylor, I. F., Watson, W. H., 1975: Glaucolide-D and glaucolide-E, two new germacranolides from *Vernonia uniflora*. − Rev. Latinoam. Quim. **6**: 191−195.

− − Adams, T. W., Watson, W. H., Jones Jr., S. B., 1976: Glaucolide G, a new germacranolide sesquiterpene lactone from *Vernonia leiocarpa* (*Compositae*). − Rev. Latinoam. Quim. **7**: 111−113.

Bevelle, C. A., Handy, G. A., Segal, R. A., Cordell, G. A., Farnsworth, N. R., 1981: Isocentratherin, a cytotoxic germacranolide from *Centratherum punctatum* (*Compositae*). − Phytochemistry **20**: 1605−1607.

Bohlmann, F., Ates (Gören), N., Jakupovic, J., 1983 a: Hirsutinolides from South African *Vernonia* species. − Phytochemistry **22**: 1159−1162.

− − − King, R. M., Robinson, H., 1984 a: Guaianolides from *Elephantopus carolinianus*. − Phytochemistry **23**: 1180−1181.

− Brindöpke, G., Rastogi, R. C., 1978 a: A new type of germacranolide from *Vernonia* species. − Phytochemistry **17**: 475−482.

− Balbaa, S. I., Halim, A. F., Halaweish, F. T., 1981 a: A terpene-coumarin derivative from *Ethulia conyzoides*. − Phytochemistry **20**: 177.

− Czerson, H., 1978 b: Neue Guajanolid-Derivate aus *Erlangea inyangana*. − Phytochemistry **17**: 568−570.

− − 1978 c: A new glaucolide derivative from *Erlangea remifolia*. − Phytochemistry **17**: 1190−1191.

− Gupta, R. K., Jakupovic, J., King, R. M., Robinson, H., 1980 a: Über das erste Sesquiterpenlacton mit einer Allengruppe aus *Vernonia cotoneaster*. − Liebigs Ann. Chem. 1904−1906.

− − − − 1981 b: Three germacranolides and other constituents from *Eremanthus* species. − Phytochemistry **20**: 1609−1612.

− Jakupovic, J., Gupta, R. K., King, R. M., Robinson, H., 1981 c: Allenic germacranolides, bourbonene derived lactones and other constituents from *Vernonia* species. − Phytochemistry **20**: 473−480.

− − King, R. M., Robinson, H., 1981 d: New germacranolides, guaianolides and rearranged guaianolides from *Lasiolaena santosii*. − Phytochemistry **20**: 1613−1622.

− Mahanta, P. K., Dutta, L. N., 1979: Weitere Hirsutinolide aus *Vernonia* Arten. − Phytochemistry **18**: 289−291.

− Müller, L., Gupta, R. K., King, R. M., Robinson, H., 1981 e: Hirsutinolide from *Vernonia* species. − Phytochemistry **20**: 2233−2237.

− − King, R. M., Robinson, H., 1981 f: A guaianolide and other constituents from *Lychnophora* species. − Phytochemistry **20**: 1149−1151.

− Scheidges, C., Misra, L. N., Jakupovic, J., 1984 b: Further glaucolides from South African *Vernonia* species. − Phytochemistry **23**: 1795−1798.

− Singh, P., Borthakur, N., Jakupovic, J., 1981 g: Three bourbonenolides and other sesquiterpene lactones from two *Vernonia* species. − Phytochemistry **20**: 2379−2382.

− − King, R. M., Robinson, H., 1982 a: New guaianolides from *Pseudostifftia kingii*. − Phytochemistry **21**: 1171−1172.

− − Robinson, H., King, R. M., 1982 b: Epi-Ilicic acid from *Alcantara ekmaniana*. − Phytochemistry **21**: 456−457.

− − Zdero, C., Ruhe, A., King, R. M., Robinson, H., 1982 c: Furanoheliangolides from two *Eremanthus* species and from *Chresta sphaerocephala*. − Phytochemistry **21**: 1669−1673.

− Wallmeyer, M., Jakupovic, J., 1982 d: Glaucolide from *Vernonia staehelinoides*. − Phytochemistry **21**: 1445−1447.

− − King, R. M., Robinson, H., 1982 e: Germacranolides from *Piptolepis leptospermoides*. − Phytochemistry **21**: 1439−1441.

– ZDERO, C., 1977 a: Inhaltsstoffe aus *Vernonia*-Arten. – Phytochemistry **16**: 778 – 779.
– – 1976 a: Curcumen-Derivat aus *Elephantopus mollis* HBK. – Chem. Ber. **109**: 3956 – 3957.
– – 1977 b: Neuartige Cumarin-Derivate aus *Ethulia conyzoides*. – Phytochemistry **16**: 1092 – 1095.
– – 1977 c: Neue 5-Methylcumarine und -chromene aus *Erlangea rogersii* S. MOORE. – Chem. Ber. **110**: 1755 – 1758.
– – 1977 d: Neue 5-Alkylcumarine and Chromone aus *Bothriocline laxa*. – Phytochemistry **16**: 1261 – 1263.
– – 1982 f: Glaucolides and other constituents from South African *Vernonia* species. – Phytochemistry **21**: 2263 – 2267.
– – KING, R. M., ROBINSON, H., 1979 a: Neue Sesquiterpenlactone aus *Stokesia laevis*. – Phytochemistry **18**: 987 – 989.
– – ROBINSON, H., KING, R. M., 1980 b: Caryophyllene derivatives and a heliangolide from *Lychnophora* species. – Phytochemistry **19**: 2381 – 2385.
– – – – 1980 c: A germacranolide from *Mattfeldanthus nobilis*. – Phytochemistry **19**: 2473 – 2474.
– – KING, R. M., ROBINSON, H., 1980 d: Sesquiterpene lactones from *Eremanthus* species. – Phytochemistry **19**: 2663 – 2668.
– – – – 1980 e: Seven guaianolides from the tribe *Vernonieae*. – Phytochemistry **19**: 2669 – 2673.
– – – – 1981 f: Two hirsutinolides and a germacranolide from *Chresta sphaerocephala*. – Phytochemistry **20**: 518 – 519.
– – – – 1981 i: Germacranolides from *Piptolepis ericoides* and *Vanillosmopsis* species. – Phytochemistry **20**: 731 – 734.
– – – – 1982 g: Hirsutinolides and other sesquiterpene lactones from *Vernonia* species. – Phytochemistry **20**: 695 – 699.
– – – – 1982 h: Germacranolides from *Stilpnopappus* species. – Phytochemistry **21**: 1045 – 1048.
– – – – 1983 b: Further hirsutinolides from *Vernonia polyanthes*. – Phytochemistry **22**: 2863 – 2864.
– – – – 1983 c: Ein neuer Typ von Sesquiterpenlacton aus *Lychnophora salicifolia*. – Liebig Ann. Chem. 1455 – 1458.
– – – – 1982 i: β-Humulene derivatives including a sesquiterpene with a rearranged carbon skeleton from *Lychnophora columnaris*. – Phytochemistry **21**: 685 – 689.
– – – – 1982 k: Germacranolides from *Lychnophora* species. – Phytochemistry **21**: 1087 – 1091.
CATALAN, C. A. N., DE IGLESIAS, D. I. A., KAVKA, J., SOSA, V. A., HERZ, W., 1986: Sesquiterpene lactones and other constituents of *Vernonia mollissima* and *Vernonia squamulosa*. – J. Nat. Prod. **49**: 351 – 351.
– LEGNAME, P. R., CRIST, B. C., DE IGLESIAS, D. I. A., 1985: Glaucolides from *Vernonia fulta*. – Phytochemistry **24**: 2113 – 2115.
CORBELLA, A., GARIBOLDI, P., JOMMI, G., FERRARI, G., 1974: Structure and abolsute stereochemistry of vanillosmin, a guaianolide from *Vanillosmopsis erythropappa*. – Phytochemistry **13**: 459 – 465.
COWALL, P. L., CASSADY, J. M., CHANG, C., KOZLOWSKI, J. F., 1981: Isolation and structure determination of piptocarphins A – F, cytotoxic germacranolide lactones from *Piptocarpha chontalensis*. – J. Org. Chem. **46**: 1108 – 1114.
COX, P. J., SIM, G. A., 1975: Sesquiterpenoids 19. X-ray crystallographic determination of the stereochemistry and conformation of the germacranolide glaucolide A. – J. Chem. Soc. Perkin 2: 455 – 458.
DAS, M. L., MAHATO, S. B., 1983: Triterpenoids. – Phytochemistry **22**: 1071 – 1095.

40   F. Bohlmann & J. Jakupovic:

De Silva, L. B., Herath, W. H. M. W., Jennings, R. C., Mahendran, M., Wannigama, G. E., 1982: A new sesquiterpene lactone from *Elephantopus scaber*. – Phytochemistry **21**: 1173–1175.

Dominguez, X. A., Cano, G., Sanchez, H., Velasquez, G., Ellmauerer, E., Jakupovic, J., 1986: A further glaucolide from *Vernonia erdverbengii*. – J. Nat. Prod. **49**: 704–705.

Fischer, N. H., Olivier, E. J., Fischer, H. D., 1979: The Biogenesis and Chemistry of Sesquiterpene Lactones. – Prog. Chem. Org. Nat. Prod. **38**: 47–390.

Ganjian, I., Kubo, I., Fludzinski, P., 1983: Insect antifeedant elemanolide lactones from *Vernonia amygdalina*. – Phytochemistry **22**: 2525–2526.

Gershenzon, J., Pfeil, R. M., Liu, Y. L., Mabry, T. J., Turner, B. L., 1984: Sesquiterpene lactones from two newly-described species of *Vernonia: V. jonesii* and *V. pooleae*. – Phytochemistry **23**: 777–780.

Govindachari, T. R., Sidhaya, R. R., Viswanathan, N., 1970: Deoxyelephantopin, a new sesquiterpene from *Elephantopus scaber*. – Ind. J. Chem. **8**: 762.

– Viswanathan, N., Führer, H., 1972: Isodeoxyelephantopin, a new germacranolide from *Elephantopus scaber*. – Ind. J. Chem. **10**: 272–273.

Harborne, J. B., Williams, C. A., 1977: The biology and chemistry of the *Compositae*. – London: Academic Press.

Herz, W., Goedken, V. L., 1982: Structure of goyazensolide and its congeners. – J. Org. Chem. **47**: 2798–2800.

– Govindan, S. V., Blount, J. F., 1981: Structures of rolandrolides and isorolandrolides, unusual germacranolides from *Rolandra fruticosa*. – J. Org. Chem. **46**: 761–765.

– Kulanthaivel, P., 1983: Piptocarphol esters from *Piptocarpha opaca*. – Phytochemistry **22**: 1286–1287.

– Kumar, N., Vichnewski, W., Blount, J. F., 1980: Cytotoxic sesquiterpene lactones of *Eremanthus incanus* and *Heterocoma albida*. Crystal structure and sterochemistry of eregoyazin. – J. Org. Chem. **45**: 2503–2506.

Jakupovic, J., Banerjee, S., Castro, V., Bohlmann, F., Schuster, A., Msonthi, J. D., Keeley, S., 1986 a: Poskeanolide, a seco-germacranolide and other sesquiterpene lactones from *Vernonia* species. – Phytochemistry **25**: 1359–1364.

– Baruah, R. N., Chau-Thi, T. V., Bohlmann, F., Msonthi, J. D., Schmeda-Hirschmann, G., 1986 b: New vernolepin derivatives from *Vernonia glabra*. – Planta Med. **52**: 378–380.

– Boeker, R., Schuster, A., Bohlmann, F., Jones, S. B., 1987 a: Further guaianolides and 5-alkylcoumarins from *Gutenbergia* and *Bothriocline* species. – Phytochemistry **26**: 1069–1075.

– Gage, D. A., Bohlmann, F., Mabry, T. J., 1986 c: Sesquiterpene lactones from *Vernonia marginata*. – Phytochemistry **25**: 1179–1183.

– Jia, Y., Zdero, C., Warning, U., Bohlmann, F., Jones, S. B., 1987 b: Germacranolides from *Elephantopus* species. – Phytochemistry **26**: 1467–1469.

– Schmeda-Hirschmann, G., Schuster, A., Zdero, C., Bohlmann, F., King, R. M., Robinson, H., Pickardt, J., 1986 d: Hirsutinolides, glaucolides and sesquiterpene lactone from *Vernonia* species. – Phytochemistry **25**: 145–158.

– Zdero, C., Boeker, R., Warning, U., Bohlmann, F., Jones, S. B., 1987 c: Vernocistifolide und andere Sesquiterpenlactone aus *Vernonia* und verwandten Arten. – Liebigs Ann. Chem. 111–123.

Jones, S. B., 1981 a: Synopsis and pollen morphology of *Vernonia* (*Compositae: Vernonieae*) in the New World. – Rhodora **81**: 425–447.

– 1981 b: Synoptic classification and pollen morphology of *Vernonia* (*Compositae: Vernonieae*) in the Old World. – Rhodora **81**: 59–75.

Kisiel, W., 1975: Phytochemical investigations of *Vernonia flexuosa* 2. Vernoflexuoside and vernoflexin, new sesquiterpene lactones. – Pol. J. Pharmacol. Pharm. **27**: 461–467.

KUPCHAN, S. M., AYNECHI, Y., CASSADY, J. M., McPHAIL, A. T., SIM, G. A., SCHNOES, H. K., BURLINGAME, A. L., 1966: The isolation and structural elucidation of two novel sesquiterpenoid tumor inhibitors from *Elephantopus elatus*. – J. Am. Chem. Soc. **88**: 3674–3676.

– HEMINGWAY, R. J., KARIM, A., WERNER, D., 1969: Tumor inhibitors. Vernodalin and vernomygdin, two new cytotoxic sesquiterpene lactones from *Vernonia amygdalina* DEL. – J. Org. Chem. **34**: 3908–3911.

– – WERNER, D., KARIM, A., McPHAIL, A. T., SIM, G. A., 1968: Vernolepin, a novel elemanolide dilactone tumor inhibitor from *Vernonia hymenolepis*. – J. Am. Chem. Soc. **90**: 3596–3597.

LEE, K. H., FURUKAWA, H., KOZUKA, M., HUANG, H. C., LUHAN, P. A., McPHAIL, A. T., 1973: Molephantin, a novel cytotoxic germacranolide from *Elephantopus mollis*. X-ray crystal structure. – Chem. Commun. 476–477.

– IBUKA, T., HUANG, H. C., HARRIS, D. L., 1975: Antitumor agents 14. Molephantinin, a new potent antitumor sesquiterpene lactone from *Elephantopus mollis*. – J. Pharm. Sci. **64**: 1077–1978.

LE QUESNE, P. W., LEVERY, S. B., MENACHERY, M. D., BRENNAN, T. F., RAFFAUF, R. F., 1978: Antitumor plants. 6. Novel modified germacranolides and other constituents of *Eremanthus elaeagnus* SCHULTZ-BIP. (*Compositae*). – J. Chem. Soc. Perkin Trans. **1**: 1572–1580.

– MENACHERY, M. D., PASTORE, M. P., KELLEY, C. J., BRENNAN, T. F., ONAN, K. D., RAFFAUF, R. F., 1982: Antitumor plants. 12. Further sesquiterpenoid constituents of *Lychnophora affinis* GARDN. (*Compositae*). X-ray structure analysis of lychnophorolide A. – J. Org. Chem. **47**: 1519–1521.

MABRY, T. J., ABDEL-BASET, Z., PADOLINA, W., JONES, S. B., 1975: Systematic implications of flavonoids and sesquiterpene lactones in species of *Vernonia*. – Bioch. System **2**: 185–192.

MAHMOUD, Z. F., SARG, T. M., AMER, M. E., KHAFAGY, S. M., BOHLMANN, F., 1980: A 5-methylcoumarin glucoside from *Ethulia conyzoides*. – Phytochemistry **19**: 2029–2030.

MALDONADO, J. E., MARTINEZ, R., MARTINEZ, V. M., 1980: Glaucolides D and E of *Vernonia liatroides* DL. – Rev. Latinoam. Quim. **111**: 58–59.

MANCHAND, P. S., TODARO, L. J., CORDELL, G. A., SOEJARTO, D. D., 1983: Isocentratherin. – J. Org. Chem. **48**: 4388–4389.

McPHAIL, A. T., MILLER, R. W., MOMPON, B., TOUBIANA, R., 1975: Structure and stereochemistry of vernodesmine, a novel phenyl bearing sesquiterpene lactone from *Vernonia pectoralis* BAKER. – Tetr. Lett. 3675–3676.

– ONAN, K. D., LEE, K. H., IBUKA, T., KOZUKA, M., SHING, T., HUANG, H. C., 1974: Structure and stereochemistry of the epoxide of phantomolin, a novel cytotoxic sesquiterpene lactone from *Elephantopus mollis*. – Tetr. Lett. 2739–2741.

MOMPON, B., HO, C. M., TOUBIANA, R., 1973: Sesquiterpenoid lactones. 6. Structure of pectorolide, a new sesquiterpenoid lactone, isolated from *Vernonia pectoralis*. – Compt. Rend. Hebd. Séances Acad. Sci. Ser. C **276**: 1799–1801.

– MASSIOT, G., TOUBIANA, R., 1974: Structure of subluteolide, a new guaianolide isolated from *Vernonia sublutea* (*Compositae*). – Compt. Rend. Hebd. Séances Acad. Sci. Ser. C **279**: 907–909.

– TOUBIANA, R., 1976: Lactones sesquiterpenique du *Vernonia pectoralis* BAKER (Composées). Stereochimie du pectorolide, et structure des vernopectolides-A et B. – Tetrahedron **32**: 2545–2548.

– – 1977: Configuration du subluteolide; nouveau guaianolide isolé du *Vernonia sublutea* SCOTT ALLIOT (Composées). – Tetrahedron **33**: 2199–2203.

Narain, N. K., 1977: Fasciculide-A, a new sesquiterpene lactone from *Vernonia fasciculata*. Spectrosc. Lett. **10**: 991−1000.

− 1978: Spectroscopic studies of fasciculide-B, a new sesquiterpene lactone from the leaves of *Vernonia fasciculata* Michx. − Spectrosc. Lett. **11**: 267−273.

Ohno, N., McCormick, S., Mabry, T. J., 1979: Centratherin, a new germacranolide from *Centratherum punctatum*. − Phytochemistry **18**: 681−682.

Padolina, W. G., Nakatani, N., Yoshioka, H., Mabry, T. J., Monti, S. A., 1974a: Marginatin, a new germacranolide from *Vernonia* species. − Phytochemistry **13**: 2225−2229.

− Yoshioka, H., Nakatani, N., Mabry, T. J., Monti, S. A., Davis, R. E., Cox, P. J., Sim, G. A., Watson, W. H., Wu, I. B., 1974b: Glaucolide-A and -B, new germacranolide-type sesquiterpene lactones from *Vernonia* (*Compositae*). − Tetrahedron **30**: 1161−1170.

Raffauf, R. F., Huang, P. K. C., Le Quesne, P. W., Levery, S. B., Brennan, T. F., 1975: Eremantholide A, a novel tumor-inhibiting compound from *Eremanthus elaeagnus* Schultz-Bip. − J. Am. Chem. Soc. **97**: 6884−6886.

Robinson, H., Bohlmann, F., King, R. M., 1980: Chemosystematic notes on the *Asteraceae*. 3.Natural subdivision of the *Vernonieae*. − Phytologia **46**: 421−436.

Rustaiyan, A., Nazarians, L., Bohlmann, F., 1980a: Germacranolides from *Erlangea cordifolia*. − Phytochemistry **19**: 1230−1232.

− − − 1980b: Two new 5-methylcoumarins from *Erlangea fusca*. − Phytochemistry **19**: 1254−1255.

− Niknejad, A., Watson, W. H., Zabel, V., Mabry, T. J., Yabuta, G., Jones, S. B. Jr., 1978: Dihydroelephantopin, a new tumor inhibitor from *Elephantopus tomentosus* L. (*Compositae*). − Rev. Latinoam. Quim. **9**: 200−201.

Shukla, V. S., Dutta, S. C., Baruah, R. N., Sharma, R. P., Thyagarajan, G., Herz, W., Kumar, N., Watanabe, K., Blount, J. F., 1982: New 5-methylcoumarins from *Ethulia conyzoides*. − Phytochemistry **21**: 1725−1731.

Toubiana, R., 1969: Structure de l'hydroxyvernolide, nouvel ester sesquiterpenique isolé du *Vernonia colorata* Drake (Composée). − Compt. Rend. Hebd. Séances Acad. Sci. Ser. C **268**: 82−85.

− Gaudemer, A., 1967: Structure du vernolide, nouvel ester sesquiterpenique isolé de *Vernonia colorata*. − Tet. Lett. 1333−1336.

− Mompon, B., Chi Man Ho, Toubiana, M. J., 1975: Isolement du vernodalin et du vernolepin à partir de *Vernonia guineensis:* authenticité du squelette elemane. − Phytochemistry **24**: 775−778.

− Toubiana, M. J., Das, B. C., 1972: Structure de confertolide, nouveau germacranolide isolé de *Vernonia conferta* (Composée). − Tet. Lett. 207−210.

Vichnewski, W., Gilbert, B., 1972: Schistosomicidal sesquiterpene lactones from *Eremanthus elaeagnus*. − Phytochemistry **11**: 2563.

− Lins, A. P., Herz, W., Murari, R., 1980: Lychnopholic acid and its acetate from *Lychnophora martiana*. − Phytochemistry **19**: 685−686.

− − Lopes, J. N. C., Filho, D. D. S., Herz, W., 1976a: 15-Deoxygoyazensolide, a new heliangolide from *Vanillosmopsis erythropappa*. − Phytochemistry **15**: 1775−1776.

− Sarti, S. J., Gilbert, B., Herz, W., 1976b: Goyazensolide, a schistosomicidal heliangolide from *Eremanthus goyazensis*. − Phytochemistry **15**: 191−193.

− Welbaneida, F., Machado, L., Rabi, J. A., Murari, R., Herz, W., 1977: Eregoyazin and eregoyazidin, two new guaianolides from *Eremanthus goyazensis*. − J. Org. Chem. **42**: 3910−3913.

Warning, U., Jakupovic, J., Bohlmann, F., Jones, S. B., 1987: Scorpiolid, ein neuer Sesquiterpenlacton-Typ aus *Vernonia scorpioides*. − Liebigs Ann. Chem. 467−468.

ZABEL, V., WATSON, W. H., MABRY, T. J., PADOLINA, W. G., 1980: Dihydroglaucolide-C, $C_{21}H_{28}O_8$. — Acta Cryst. B **36**: 3024–3027.

ZDERO, C., BOHLMANN, F., ROBINSON, H., KING, R. M., 1981: Germacranolides from *Proteopsis argentea*. — Phytochemistry **20**: 739–741.

— — 1988: Macrolide diterpene and other ent-labdanes from *Corymbium villosum*. — Phytochemistry **26**: 227–231.

— — KING, R. M., ROBINSON, H., 1986: α-Isocedrene derivatives, 5-methyl coumarins and other constituents from the subtribe *Nassauviinae* of the *Compositae*. — Phytochemistry **25**: 2873–2882.

Address of the author: Prof. Dr D. F. BOHLMANN and Dr J. JAKUPOVIC, Technical University, Department of Organic Chemistry, Strasse des 17. Juni 135, D-1000 Berlin 12.

Pl. Syst. Evol. [Suppl. 4], 45 – 66 (1990)

A preliminary cladistic analysis of the genus *Vernonia* (*Vernonieae: Asteraceae*)

STERLING C. KEELEY and BILLIE L. TURNER

Received March 16, 1988

Key words: Angiosperms, *Asteraceae, Vernonieae, Vernonia.* – Cladistics, phylogeny.

Abstract: A cladistic study of the sections and subsections of the genus *Vernonia* was performed using 32 characters. The resulting cladograms indicate that New and Old World taxa are largely distinct, but differences in topology of equally parsimonious trees indicate that the level and position of connections is unclear. The yellow-flowered trinervate-leaved species of Madagascar separate out as distinct from other Old World taxa as do the large-headed leafy-bracted species of the New World *Leiboldia* group. Problems are encountered with the placement of the New World subsect. *Eremosis/Critoniopsis* and the Old World subsectt. *Strobocalyx* and *Urceolatae*. Our analyses support the overall synoptic treatment of JONES and JEFFREY, however there are areas of phylogenetic and taxonomic difficulty which need further study. Other non-morphological characters are required to resolve problematic relationships and provide a taxonomic framework reflective of phylogenetic relationships.

The tribe *Vernonieae* is one of the most poorly known of the family *Asteraceae* despite its worldwide distribution and large number of species. 1500 taxa in over 70 genera are currently recognized (Table 1). However, the status and position of many members remains muddled due to lack of information on important aspects of the biology of these taxa, e.g., chromosome numbers, chemistry, and morphological variability. The distribution of taxa within genera is anomalous within the tribe *Vernonieae*: most genera are monotypic or have fewer than 10 species. The vast majority of species, approximately 1000, are found within the "core" genus, *Vernonia*.

Current tribal arrangements are based largely on the treatment of BENTHAM & HOOKER (1873), but there is much dispute as to appropriate subtribal groupings (e.g., ROBINSON & BRETTEL 1973, JEFFREY 1988, TURNER 1981, ROBINSON & FUNK 1987). One of the most significant problems in this regard, centers around the concept and delimitation of the core genus *Vernonia* (e.g., GLEASON 1906; EKMAN 1914; BLAKE 1936; ROBINSON 1977, 1980, 1984; ROBINSON & BRETTEL 1973; ROBINSON & FUNK 1987; JUNK 1977; TURNER 1981; JEFFREY 1988). Information has been gained in a number of areas, i.e., from palynology (KINGHAM 1976; BOLICK 1978; KEELEY & JONES 1977a, 1979), chemistry (MABRY & al. 1975; ROBINSON & al. 1980, KING & JONES 1982, SEAMAN 1982, BOHLMAN & JAKUPOVIC 1990), cytology

Table 1. Currently recognized genera of the *Vernonieae*, number of taxa, and geographical location. Base chromosome number in parentheses where known

	No. of species	Reference	Geographical distribution
Acanthodesmos ADAMS	1	WILLIS 1985	Jamaica
Adenoon DALZ.	1	WILLIS 1985	Indo-Malaysia (10)
Aedesia EATON	2	COILE 1981, WILLIS 1985	W. Africa (10)
Ageratinastrum MATTF.	5	WILLIS 1985	Trop. Africa
Albertinia SPRENG.	1	WILLIS 1985	Brazil
Alcantara GLAZIOU	1	ROBINSON 1981, WILLIS 1985	Brazil
Argyrovernonia MACLEISH	1	MACLEISH 1984 b	Brazil
Bishopalea H. ROBINSON	1	ROBINSON 1981	Brazil
Blanchetia DC.	1	MACLEISH 1984 A	Brazil
Bothriocline OLIV. ex BENTH.	40	JEFFREY 1988, pers. comm.; WILD & POPE 1977	Trop. Africa (9, 10)
Brachythrix WILD & POPE	5	WILD 1978	Trop. Africa
Camchaya GAGNEP.	5	KITAMURA 1968	Indochina (10)
Centauropsis BOJ. ex DC.	8	JEFFREY 1988, pers. comm.; HUMBERT 1960, WILLIS 1985	Madagascar
Centratherum CASS.	3	KIRKMAN 1981	New and Old World tropics (8, 9)
Chresta VELL.	4	MACLEISH 1985 b, 1987	Brazil
Chronopappus DC.	1	WILLIS 1985	Brazil
Critoniopsis SCH. BIP.	26	ROBINSON 1980	S. America
Decastylocarpus HUMBERT	1	HUMBERT 1960, WILLIS 1985	Madagascar
Dewildemannia O. HOFFM.	3	WILD & POPE 1977	Trop. Africa
Diaphractanthus HUMBERT	1	HUMBERT 1960, WILLIS 1985	Madagascar
Dipterocypsela BLAKE	1	WILLIS 1985	Columbia
Distephanus CASS.	24	ROBINSON & KAHN 1986	Madagascar, Trop. Africa
Ekmania GLEASON	1	WILLIS 1985	Cuba
Elephantopus L.	27	CLONTS 1972, BUSEY 1975	New and Old World tropics (11, 13)
Episcothamnus H. ROBINSON	1	ROBINSON 1981	Brazil
Eremanthus LESS.	18	MACLEISH 1987	Brazil (10)
Erlangea SCH. BIP.	9	JEFFREY 1988	Trop. Africa (10)
Ethulia L.	20	JEFFREY 1988, pers. comm.; WILD & POPE 1977	Trop. Africa (10)
Glaziovianthus G. BARROSO	2	MACLEISH 1985 a	Brazil
Gorceixia BAKER	1	WILLIS 1985	Brasil
Gossweilera S. MOORE	2	WILD & POPE 1977	Angola
Gutenbergia SCH. BIP.	20	JEFFREY 1988, pers. comm.; WILLIS 1985	Trop. Africa (10)
Haplostephium MART. ex DC.	3	WILLIS 1985	Brasil
Harleya BLAKE	1	WILLIS 1985	Mexico, C. America
Herderia CASS.	1	WILLIS 1985	Trop. Africa
Heterocoma DC.	2	WILLIS 1985	Brazil
Heterocypsela H. ROBINSON	1	ROBINSON 1979	Brazil
Hoplophyllum DC.	2	JEFFREY 1988, pers. comm.	S. Trop. Africa
Hystrichophora MATTF.	1	JEFFREY 1988, pers. comm.; WILLIS 1985	E. Trop. Africa

Table 1 continued

	No. of species	Reference	Geographical distribution
Iodocephalus Thorel ex Gagnep.	3	Willis 1985	S.E. Asia
Kinghamia C. Jeffrey	4	Jeffrey 1987, pers. comm.	E. Trop. Africa
Lachnorhiza A. Rich.	1	Willis 1985	Cuba
Lamprachaenium Benth.	1	Willis 1985	India (9)
Leiboldia Schlechtd. ex Gleason	1	Robinson & Funk 1987	Mexico, C. America (19)
Lepidonia Blake	7	Robinson & Funk 1987	Mexico (19)
Lychnophora Mart.	11	Coile 1981	Brazil (17)
Mattfeldanthus H. Robinson	1	Robinson & King 1979	Brazil
Msuata O. Hoffm.	1	Willis 1985	Trop. Africa
Muschleria S. Moore	1	Wild & Pope 1977	Trop. Africa (10)
Neurolakis Mattf.	1	Willis 1985	Trop. Africa
Oiospermum Less.	1	Willis 1985	Brazil
Oliganthes Cass.	1	Aristeguieta 1963, Stutts 1981	Madagascar
Omphalopappus O. Hoffm.	1	Wild & Pope 1977	Angola
Pacourina Aubl.	1	Willis 1985	C. & S. America
Paralychnophora MacLeish	3	MacLeish 1984 b	Brazil
Phyllocephalum Blume	3	Kirkman 1981	Old World Tropics
Piptocarpha R. Br.	43	Smith 1982, 1984	S. America (17)
Piptocoma Cass.	3	Stutts & Muir 1981	S. America
Piptolepis Benth.	8	Willis 1985	Brazil
Pithecoseris Mart.	1	Willis 1985	Brazil
Pleurocarpaea Benth.	1	Willis 1985	Australia
Pollalesta Kunth	16	Stutts 1981	C. and S. America
Prestelia Sch. Bip.	1	MacLeish 1984 a	Brazil
Proteopsis Mart. & Zucc. ex DC.	5	Willis 1985	Brazil
Pseudelephantopus Rohr.	2	Busey 1975; Jeffrey 1988, pers. comm.	New and Old World tropics
Pseudostifftia H. Robinson	1	Robinson 1979	Brazil
Pycnocephalum DC.	3	MacLeish 1985 b	Brazil
Rastrophyllum Wild & Pope	2	Jeffrey 1988, pers. comm.	E. Africa
Rolandra Rottb.	1	Willis 1985	S. America
Sipolisia Glaziou	1	Robinson 1981	Brazil
Soaresia Sch. Bip.	1	Willis 1985	Brazil
Sparganophorus Adans.	1	Jeffrey 1988, pers. comm.	New and Old World tropics (16)
Spiracantha Kunth	1	Willis 1985	Colombia
Stephanolepis S. Moore	1	Willis 1985	Trop. Africa
Stilpnopappus Mart. ex DC.	20	Willis 1985	S. America
Stokesia L'Herit.	1	Willis 1985	S.E. North America (7)
Stramentopappus H. Robinson & V. Funk	1	Robinson & Funk 1987	Mexico (19)
Telmatophila Mart. ex Baker	1	Willis 1985	Brazil
Trichospira Kunth	1	Willis 1985	Trop. America (10)
Vernonia Schreb.	1 000	Willis 1985	New and Old World (see Table 2)

(JONES 1976, 1979 a; TURNER 1981; RABAKONANDRIANINA & CARR 1987), and morphology (ROBINSON & BRETTEL 1973; KEELEY & JONES 1977 a, b; JONES 1979 c, 1981; KEELEY 1978, 1982; JEFFREY 1988; POPE 1983; ROBINSON & FUNK 1987), but nevertheless relationships remain unclear. JONES (1979 c, 1981) provided the only comprehensive treatment of *Vernonia* at the sectional and subsectional levels (Table 2), but recent work has resulted in a number of alterations (TURNER 1981, ROBINSON 1980, ROBINSON & KAHN 1986, ROBINSON & FUNK 1987, JEFFREY 1988).

It was the purpose of this study to use cladistic methodology in an attempt to understand the relationships within the core genus *Vernonia* at the sectional and subsectional levels. The large number of taxa, their uneven representation in herbaria and the range of morphological variation within sections and subsections makes assessments only preliminary. Thus, the relationships proposed here must be regarded as working hypotheses which can, and should be revised as more and better information becomes available, especially from macromolecular studies.

Materials and methods

Relationships among the sections and subsections of *Vernonia* were analyzed using primarily morphological characters assembled from the herbarium and by familiarity of selected groups through the ongoing systematic studies by the senior author. The sectional and subsectional placements of taxa are according to JONES (1979 c, 1981) with some recent modifications as shown in Table 2. Data were obtained from type species of each section and subsection whenever possible; no attempt was made to assess the considerable variability found in some of the larger sections. Material was obtained from BM, BR, C, G, GH, K, LL, M, MO, NY, P, S, TEX, and US. Literature was used to supplement information from specimens when available material was poor, and to provide information on chromosome number, pollen type (Table 2), and chemistry. Data were analyzed using SWOFFORD's (1985) program, Phylogenetic Analysis Using Parsimony (PAUP version 2.4) for microcomputer. Cladograms presented here were rooted using the outgroup method (WATROUS & WHEELER 1981). The "unordered" option was used to avoid a priori assumptions about transformation series (MEACHAM 1984, SWOFFORD 1985). The options AD-DSEQ = CLOSEST, SWAP = GLOBAL and MULPARS were used. The data matrix is given in Appendix 1.

Character analysis

Outgroup choice. The appropriate choice of an outgroup to *Vernonia* and the *Vernonieae* has been difficult because no tribal phylogeny exists, and agreement is lacking on the constitution and delimitation of *Vernonia* (e.g., TURNER 1981, RO-BINSON & FUNK 1987, JEFFREY 1988). In addition, tribal relationships have been unclear until recently (CASSINI 1825, BENTHAM 1873, AUGIER & DUMERAC 1951, CRONQUIST 1955, CARLQUIST 1976, WAGENITZ 1976, TURNER 1977). BREMER (1987) used cladistic methodology to study the tribes and subtribes within the *Asteraceae* using a broad range of characters. He found the *Liabeae* to be the closest tribe to the *Vernonieae*. The two tribes have similar styles, stylar trichomes and overall pollen morphology in addition to other features which support the clade which includes the *Liabeae* and the *Vernonieae*, and, more distantly, the tribes *Eremothamneae* and *Lactuceae*. In addition, a phylogenetic study based on chloroplast DNA restriction site mutations in the *Asteraceae*, also places the *Liabeae* (*Liabum* and *Cacosmia*) next to the *Vernonieae* (*Vernonia, Stokesia, Lychnophora, Piptocarpha*) (JANSEN & PALMER pers. comm.).

Table 2. Synoptic classification of *Vernonia*, pollen types and base chromosome numbers. Based on JONES (1979c, 1981) and chromosome data available since 1981 (see text). [1] Modified by TURNER (1981) and ROBINSON & FUNK (1987); see Discussion. [2] Partially modified by ROBINSON (1980)

	Pollen type	Base chromosome no.
New World		
Subg. *Vernonia*		
sect. *Leiboldia*[1]	A	19
sect. *Hololepis*	A	–
sect. *Vernonia*		
subsect. *Noveboracenses*	A	17
subsect. *Eremosis*[2]	A	10, 17?
subsect. *Polyanthes*	A	10, 17?
subsect. *Buxifoliae*	A	–
subsect. *Chamaedrys*	A, B	17
subsect. *Stenocephalum*	C	12, 17?
subsect. *Nudiflorae*	A, B	17
subsect. *Scorpioides*	B, C (A)	10, 14, 15, 17
Old World		
Subg. *Orbisvestus*		
sect. *Orbisvestus*	A (B)	9, 10
subsect. *Orbisvestus*	A (B)	9, 10
subsect. *Strobocalys*	A (B)	10
subsect. *Gongrothamnus*	A	9
subsect. *Pawekianae*	A	9
subsect. *Hilliardianae*	A	9
subsect. *Urceolatae*	A	9?
subsect. *Tubinellae*	A	9?
subsect. *Distephanus*	A	10
subsect. *Centrapalus*	A	9
subsect. *Stengelia*	C	10
sect. *Tephrodes*	E	9, 10
subsect. *Tephrodes*	E	9
subsect. *Lepidella*	E	9, 10
subsect. *Oocephalae*	E	9?
subsect. *Glutinosae*	E	9?
subsect. *Bechium*	E	9?
sect. *Azureae*		10

For our analysis we chose the *Liabeae* as the most appropriate outgroup. This choice was based on several factors: the close relationships between *Vernonieae* and *Liabeae* shown by the work of BREMER (1987) and JANSEN & PALMER (pers. comm.) discussed above, the recent recognition of the *Liabeae* as a distinct, but closely allied tribe to the *Vernonieae* where *Liabeae* genera have traditionally been placed (ROBINSON 1983), and on our work which showed that morphological data currently available were not sufficiently powerful to resolve generic relationships within the *Vernonieae*. Thus, there was no clear logical choice of closest sister group within the *Vernonieae*, and it did not make sense to us to arbitrarily choose a taxon without knowing more about its relationships to other taxa within the tribe. Current disagreements as to generic status of taxa traditionally placed within *Vernonia* (TURNER 1981, ROBINSON & KAHN 1986, ROBINSON & FUNK 1987), and JEFFREY's (1988) suggestion that Old and New World elements of *Vernonia* are distinct enough to be treated as separate genera, also indicated that a more distant outgroup (the *Liabeae*) was required in order to evaluate relationships within the many and geographically disparate elements of *Vernonia*.

Leaf characters. 1. Phyllotaxy. Opposite leaves are relatively uncommon in the *Vernonieae*, but are a regular feature of the *Liabeae*. Some S. American taxa have opposite to sub-opposite leaves (ROBINSON 1980, KEELEY & DIAZ unpubl.). This group comprised of half a dozen taxa has not yet been placed within the currently available taxonomic framework, but are related to subsect. *Eremosis* (*Critoniopsis*) (CUATRECASAS 1956, ROBINSON 1980, KEELEY & DIAZ unpubl.). These taxa were deleted from the final analysis, but would appear with other taxa in the above subgroup.

2. Venation. Most members of the *Vernonieae* have pinnately veined leaves, but members of Old World subsectt. *Gongrothamnus*, *Distephanus*, and *Glutinosae* are trinervate. In addition, *Vernonia peduculata* DC. of the New world sect. *Hololepis* has leafy trinervate bracts subtending the heads. These leafy bracts are similar to the cauline foliage although the lower leaves are pinnately veined. The *Liabeae* is characterized by trinervate leaves. BREMER (1987) notes that trinervate leaves appear sporadically in a number of genera in several tribes, including some members of the *Barnedesiinae*, a subtribe of the *Mutisieae*, and may be an ancestral character within the family. The trinervate condition appears to be primitive and largely restricted to Old World taxa in the *Vernonieae*.

3. Uniseriate trichomes. A wide variety of uniseriate trichomes exists in *Vernonia* and the *Vernonieae* and can be of taxonomic significance (FAUST & JONES 1973, POPE 1983, JEFFREY 1988). There is a tendency for Old World taxa to have long multicellular trichomes of a type not found in the New World. Likewise, stellate trichomes predominate in a number of Andean genera such as *Vernonia*, *Piptocarpha*, *Pollalesta*, and *Lychnophora*. This character was coded as multistate, but no presumption of evolutionary pathway, i.e., simple to t-shaped to stellate, for example was made.

Floral and inflorescence characters. 4. Florets/head. There are broadly speaking 3 categories of floret number in the *Vernonieae* (Table 3). In the past it has been generally assumed that heads with 100 or more florets were probably primitive within the tribe whereas decreasing number, to the limiting case of 1 floret/head, was evolutionary more advanced. The great majority of *Vernonia* taxa have some intermediate number. Since there are a number of taxa with both very few and a great many florets, this character was analyzed as unordered to remove a priori assumptions on the direction or uniformity of evolution. Within the *Liabeae, Casosmia* has very few florets/head whereas *Liabum* was an intermediate number, other genera such as *Austroliabum* have large numbers of florets.

5. Corolla color. Corolla color in the *Vernonieae* is predominately purple,

Table 3. Characters and states used for analysis of *Vernonia* sections and subsections

1. Phyllotaxy: (A) alternate (B) opposite
2. Venation: (A) pinnate (B) trinervate
3. Uniseriate trichomes: (A) simple (B) branched (C) stellate (D) peltate (E) multicellular > 3 cells (F) none
4. Floret/head: (A) >100 (B) 10−60 (75) (C) 1−9
5. Corolla color: (A) purple (pink) (B) white (C) blue (D) yellow (E) red
6. Tube/limb ratio: (A) Tube > limb (B) Tube = limb (C) tube < limb
7. Throat/tube transition: (A) gradual (B) abrupt
8. Corolla glands: (A) present (B) absent
9. Peripheral florets: (A) regular 5-lobed (B) irregular 5-lobed
10. Capitulescence type: (A) 1-few terminal heads (B) multi-head raceme or compound corymbose/paniculate (C) simple leafy cyme (reduced)
11. Palea: (A) present (B) absent
12. Receptacle shape: (A) conical (B) convex (C) flat (D) concave
13. Receptacle pitting: (A) smooth (B) alveolate (C) sclerified projections between pits (F) pubescent
14. Receptacular bracts, series: (A) many (B) 4−6 (8) (C) 4 or fewer
15. Recept. bracts, imbrication: (A) strong (B) weak or none
16. Recept. bracts, duration: (A) persistent (B) deciduous (C) tardily or semi-deciduous
17. Recept. bract, form: (A) foliaceous (B) sub-/semi-foliaceous (C) scarious
18. Pappus members (A) 1 kind all alike (B) 2 kinds, not alike (C) none
19. Pappus heterogeneity: (A) one (B) two (C) >2 (D) zero
20. Pappus duration (any part): (A) persistent (B) deciduous (C) none
21. Pappus fusion (any part): (A) free (B) fused (C) none
22. Pappus type: (A) simple bristles scales (B) modified (C) none
23. Anther base: (A) sagittate (B) caudate (C) papillate
24. Nectary: (A) absent or poorly developed (B) present
25. Style trichomes (midpoint on style branches): (A) acute (B) rounded (C) blunt (D) lageniform (E) clavate or other
26. Style base expanded: (A) yes (B) no
27. Achene rib no.: (A) c. 10 (8−10) (B) 2−5 (7) (C) none
28. Achene trichomes: (A) uniseriate (B) multiseriate (C) zwillingshaare (D) none
29. Habit: (A) woody perennial (B) herbaceous perennial (C) annual
30. Pollen type: (A) A (B) B or C (C) D (D) E (E) F
31. Chromosome number: (A) x=9, 10 (7−14) (B) x=17, 18 (C) x=19
32. Sesquiterpene lactones: (A) New World glaucolides (B) Old World germacranolides

varying to paler shades of lavender, pink or white, but yellow, red and blue also occur. We scored each color as it occurred, without assumption as to the direction of the color change. Corolla color in the *Liabeae* is yellow or orange.

6. Tube/limb ratio. The length of the tube relative to the limb varies within several groups of *Vernonieae* and was found to be informative. This character is also found to vary within the *Liabeae*. It is difficult to presume change in direction for this character, however, since little is known of the full extent of this feature in the *Asteraceae*.

7. Throat/tube transition. During studies by the first author on Andean Vernonias it became apparent that this character could be used to identify related lineages. It was also found to vary within members of the *Vernonieae* and the

Liabeae. A strongly constricted throat is relatively uncommon, and is apparently derived.

8. Corolla glands. This character has been used in work by POPE (1983) and JEFFREY (1988) in descriptions of African *Vernonieae*. This character was scored simply as presence absence after initial attempts to characterize types were unsuccessful. Gland presence also varied in *Liabeae*. This character was used for a number of runs, but was eventually excluded from the analyses because of the large number of changes.

9. Peripheral florets. Most members of the *Vernonieae* have eradiate heads and florets with 5-lobed corollas. However, there are notable exceptions, for example in *Stokesia* or *Elephantopus*; indeed occasional *Vernonia* spp. show a tendency to elongation of lobes in the peripheral florets. This is especially so for some of the large-headed African taxa, where it appears to be a derived condition. The *Liabeae* is characterized by having true ray florrets.

10. Capitulescence type. Capitulescence type varies widely in the *Vernonieae* and has been used to establish subgeneric groups in *Vernonia* (GLEASON 1906, 1922, 1923; JONES 1979c, 1981). Given the range of variation we chose broad categories which were useful at this level (Table 3). Again, no a priori assumptions were made on the direction of change or ancestral type. *Liabeae* has terminal heads, and heads in racemes or panicles.

Receptacle characters. 11. Palea. Palea occur sporadically in the *Vernonieae* and *Liabeae*. In general, this is an uncommon condition for most tribes of the *Compositae,* as noted by BREMER (1987). Since few taxa in either the *Liabeae* or the *Vernonieae* have palea, it is most parsimonious to assume that this is a derived condition. This character was treated as "unordered" to avoid any a priori assumptions, however. Coding was changed on one run for this character in the *Liabeae* since there are paleaceous genera, but the topology was unaffected.

12. Receptacle shape. Receptacle shape varied from conical to concave, and proved to be extremely variable. A conical receptacle appeared to be primitive, but transitions to other states were difficult to interpret. We excluded this character from the final analyses because of a high level of homoplasy and the difficulty of interpretation.

13. Receptacle pitting. Pitting of the receptacle is prominent in a number of genera in the tribe, as is the occasional presence of pubescence or projections. This character showed such a range of variation that we eventually excluded it from our analyses. It is unclear how transitions should be characterized, and may be useful at lower, rather than higher, taxonomic levels.

Involucral characters. 14. Series of involucral bracts. The pluriseriate involucre is widespread in the family (BREMER 1987), but most taxa in the *Vernoniae* have between 5 – 8 series. Large numbers of bracts are found in some members of *Vernonia* and in other genera of the tribe; likewise, reduction to just a few series is also known. The three states shown (Table 3) allowed us to characterize major differences; this character was treated as "unordered" since we could not be sure of direction of change in all cases. A similar range in involucral bract number was also found for the *Liabeae*.

15. Imbrication. The involucre in *Vernonieae* is typically regularly imbricate or graduate, however variations occur, e.g. irregularly imbricate bracts in *Vernonia cinerea*. The character was treated as unordered, and proved to be highly variable.

The final runs excluded this since there was high homoplasy and changes were difficult to interpret. Variations in degree of imbrication also occur in the *Liabeae*.

16. Duration. Involucral bracts of most genera of the *Vernonieae* are persistent, but are partly or wholly deciduous in others (e.g., *Piptocarpha*). In those with persistent bracts, the involucre often reflexes at maturity exposing the achenes to air currents which presumable affect dispersal. Deciduous bracts may accomplish the same purpose, and are often found in taxa with small heads or congested capitulescences. Deciduous bracts in the *Liabeae* are present in some genera, such as *Cacosmia* but, not in others, e.g., *Liabum*. They appear to be derived in *Vernonia*.

17. Form. A range from foliaceous to scarious involucral bracts is found in the *Vernonieae*. These have been used to delimit sections and subsections (SMITH 1971, ROBINSON & FUNK 1987). These were broadly categorized for each taxon studied. It has been generally assumed (CRONQUIST 1955) that the more foliaceous bracts are primitive since they most closely resemble leaves, but this is not common in *Vernonia*. Some leafy bracts are also colored and appear to be specialized for pollinator attraction, e.g., stengelioid vernonias (SMITH 1971). We did not presume a direction of evolutionary progression, a priori, since a number of types occur and intergrade. Leafy bracts are relatively uncommon in *Vernonia*, but are found in both New and Old World sections.

Pappus characters. 18. Pappus heterogeneity. *Vernonia* taxa generally have a biseriate pappus, often with a longer inner series of bristles, and a shorter outer series, or either bristles or scales. There is much variation however, and this has been used at all taxonomic levels within the tribe. Since it is part of the dispersal system and likely to change, along with number of florets and deciduous phyllaries, we did not assume a linear evolutionary sequence.

19. Series. Although it is typical for *Vernonia* to have two series there is variation in this character. ROBINSON & FUNK (1987) have suggested that this variation extends beyond the definition of the genus *Vernonia*, and taxa which have other than two series should be excluded (see Discussion). The number of pappus series varies within the *Liabeae* as well.

20. Duration. Pappus members can be persistent or very fragile and deciduous in the *Vernonieae*. In some cases only one series, usually the inner, is deciduous, or all parts may be. This character has been used to provide taxonomic distinctions at several levels within the tribe. Variations in duration also occur in the *Liabeae*.

21. Fusion. As in the above cases, variation occurs in fusion of pappus members. Similar variations are encountered in the *Liabeae*. In general fused pappus appears derived.

22. Type. The pappus of most members of the *Vernonieae* is of bristles, scales, or both. Coroniform, or fused pappus scales, and other variations also exist. Most members of the *Liabeae* also have simple bristles or scales.

Anther and style characters. 23. Anther base. Most members of the *Vernonieae* have sagittate anther sacs, but variation occurs. Differences in the basal appendages of anthers has been considered taxonomically significant at the generic and tribal levels by nearly all early workers, and was especially emphasized by BENTHAM (1873) for tribal recognition. Variation in this feature also occurs in the *Liabeae*. However, as BREMBER (1987) noted distinctions between types are not always clear, and further study is needed. We assigned taxa to one of three general categories which were relatively easy to distinguish, but did not assume unidirectional change.

24. Nectary. This character was chosen after we began scoring taxa for enlarged style base or stylar node (character 26) described in an analysis of *Vernonieae* genera by ROBINSON & FUNK (1987). In most cases, we found that taxa with a clearly evident nectary lacked an enlarged stylar base, however there were a number of exceptions. We felt that we should score these two characters separately and found them to be useful. No polarity, however, was assigned to this character.

25. Style trichomes. Style trichomes have been used at the generic level by ROBINSON (1980). The presence of slender hairs along the style and shaft has also been used by BREMER (1987) at the tribal level. Since both the *Vernonieae* and the *Liabeae* share this feature we were interested in differences in trichome morphology which could be taxonomically useful within *Vernonia* and the tribe. Distinct trichome types were scored; no a priori assumptions as to polarity were made.

26. Style base expanded. This character was noted by ROBINSON & FUNK (1987), as described above. We found that a distinctly expanded base was notable in a number of taxa in the *Vernonieae* and lacking in others. Most genera of *Liabeae* also lacked this character, but at least one species of *Philoglossa* and one of *Sinclairia* have expanded style bases.

Cypsela characters. 27. Rib number. Rib number has been used to distinguish taxa at several levels within the *Vernonieae* (JONES 1979 c, POPE 1983, JEFFREY 1988) and varies also within the *Liabeae*. This character was highly homoplasious and appears to be useful above the sectional level. We excluded this character from the final analyses since changes were so frequent and could not be readily interpreted.

28. Trichomes. Trichomes or hairs have been useful in distinguishing among taxa of the *Vernonieae* (POPE 1983). Most *Vernonieae* have the typical "Zwillingshaare", but may have other types in addition, or may lack trichomes entirely. *Liabeae* genera appear to have the general twin-hair type of the family, or these may be absent.

Habit. 29. Habit. It is generally assumed that the shrubby habit is primitive in *Asteraceae* (CARLQUIST 1976; CRONQUIST 1977) and that annuals and trees are derived (BREMER 1987). *Vernonieae* include all life forms from annuals to trees, likewise *Liabeae* include a range but are predominately shrubby. Since it is likely that evolution has proceeded in different directions from a shrubby ancestor we did not specify a directed evolutionary sequence.

Pollen. 30. Pollen type. This character was scored according to the pollen types identified for *Vernonia* by KEELEY & JONES (1977 a, 1979). Types were classified to reflect the common patterns seen within *Vernonia* subsections. These were grouped where two types occurred. The pollen of the *Liabeae* is similar to that of the *Vernonieae* pollen (SKVARLA & al. 1977), but is distinct, and is not here considered a direct antecedent to *Vernonieae* pollen.

Chromosomes. 31. Chromosome number. Base chromosome numbers for the Old World subg. *Orbisvestus* are $x = 9$ and $x = 10$, although one recent count of $x = 7$ was reported for *Vernonia appendiculata* of Madagascar (RABAKONANDRIANINA & CARR 1987). New World taxa appear to have a base number of $x = 17$ (JONES 1979 c), but a number of other counts have been recorded (Table 2) (TURNER & al. 1967, COLEMAN 1968, JONES 1973, JANSEN & al. 1984, STROTHER 1983). Given the range in numbers it is possible that subsectt. *Scorpioides, Stenocephalum*, and *Eremosis* may be multibasic with $x = 9, 10, 14, 16, 17,$ and 18. More data are needed before this can be resolved. Chromosome counts of $n = 19$ have been reported for three New World taxa, *V. arctioides, V. callilepis,* and *V. pooleae* of sectt. *Leiboldia, Lepidonia* and *Stramentopappus* (SUNDBERG & al. 1986; TURNER 1981; and unpubl.; ROBINSON & FUNK 1987). The relationships among New World vernonias may be more reticulate than in the Old World. TURNER (1981) has postulated that this reticulation is perhaps quite ancient, having evolved from ancestral phyletic lines with $x = 10$ and $x = 9$. the $x = 19$ lines may have arisen by aneuploid reduction $2x = 20$ or possibly by ancestral amphiploidy $(19 = 9 + 10)$. Chromosomal relationships among $x = 17$ taxa may have arisen by descending aneuploidy (i.e., $n = 18 - 1$) from polyploid lines that connect back to $x = 9$ or 10 lines, which are common in Old World lineages. *Liabeae* genera reported to date (NORDENSTAM 1977, DILLON & TURNER 1982) have a base number of $x = 9$, counts of $x = 10$ and 12 have also been reported. Base numbers of $x = 9$ and 10 are common within the *Asteraceae*, occurring in almost every tribe. However, there is a smattering of chromosome numbers that suggest an ancestral base number for the family may be $x = 4$ or 5 (TURNER 1977).

Chemistry. 32. Sesquiterpene lactones. Sesquiterpene lactones are known from a number of taxa in *Vernonia*, and the *Vernonieae* (SEAMAN 1982, BOHLMANN & JAKUPOVIC 1990) and have been used at the sectional (TURNER 1981) and subtribal levels (ROBINSON & al. 1980). A current analysis of known chemistry is provided by BOHLMANN & JAKUPOVIC (1990) and was beyond the boundaries of this study. However, as summarized in SEAMAN (1982) New World *Vernonia* spp. are sufficiently distinct chemically from Old World taxa to represent two different phytogeographical lines. Only *V. jonesii* of the New World sect. *Lepidonia* has nonglaucolide germacranolides as found commonly in Old World taxa (GERSHENZON & al. 1984). The *Liabeae* does contain some germacranolides and germacrolides like the *Vernonieae* (SEAMAN 1982), however the tribe is not well studied. To avoid assumptions of evolutionary direction involving the outgroup we coded this character as missing for the *Liabeae*.

Discussion

Results of the analysis are shown in Figs. 1 and 2. They show two representative topologies among the 100 equally parsimonious trees (the maximum allowed by PAUP) which were found. In all trees the basic clades remained the same and were supported by the same synapomorphies; within each of these two topologies minor rearrangements occurred on the terminal branches. The principal difference between the two topologies shown is in the position of the clade formed by sectt. *Eremosis, Critoniopsis, Strobocalyx,* and *Urceolatae* (discussed below). The consistency index was 0.52; total homoplasy was 48%, similar to the values of $50 - 80\%$ homoplasy found in studies of other *Asteraceae* genera, (FUNK 1982, 1985; JANSEN 1985; JANSEN & al. 1987).

The analyses support the recognition of a number of monophyletic groups within *Vernonia*. These are largely those identified by JONES (1979 c, 1981), JEFFREY (1988), and others (see below). However, there were repeated parallelisms and reversals (Figs. 1 and 2) within a number of morphological characters, such as habit (29), number of florets/head (4), style trichomes (25), expanded style base (26), phyllary type (17), and series (14) which have been used at all taxonomic levels within *Vernonia*. The difference in placement of clades reflected by the two topologies

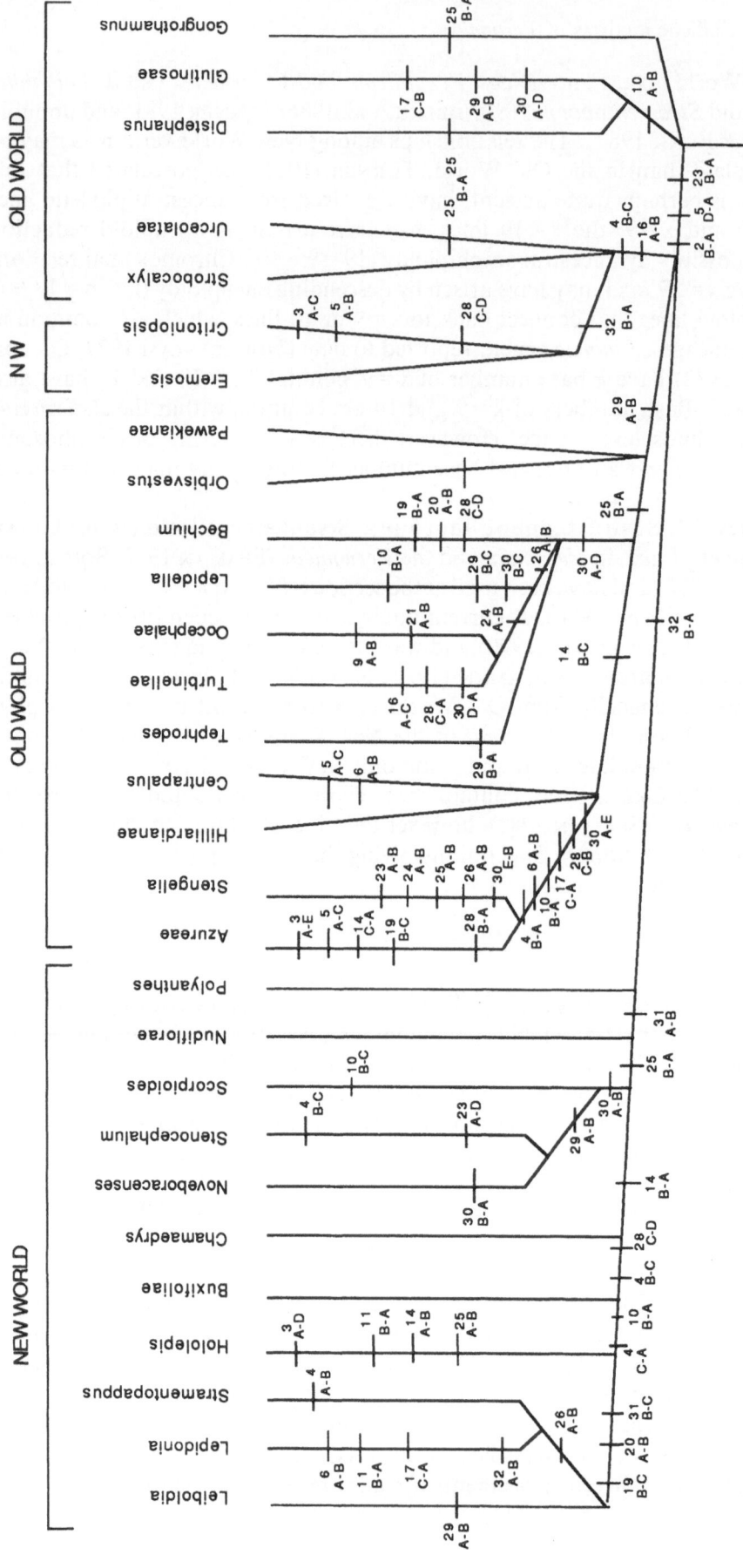

Fig. 1. Cladogram of the sections and subsections of *Vernonia*. Numbers correspond to characters and character states in Table 3, character state changes indicated by letters. Tree length 91, CI = 0.52

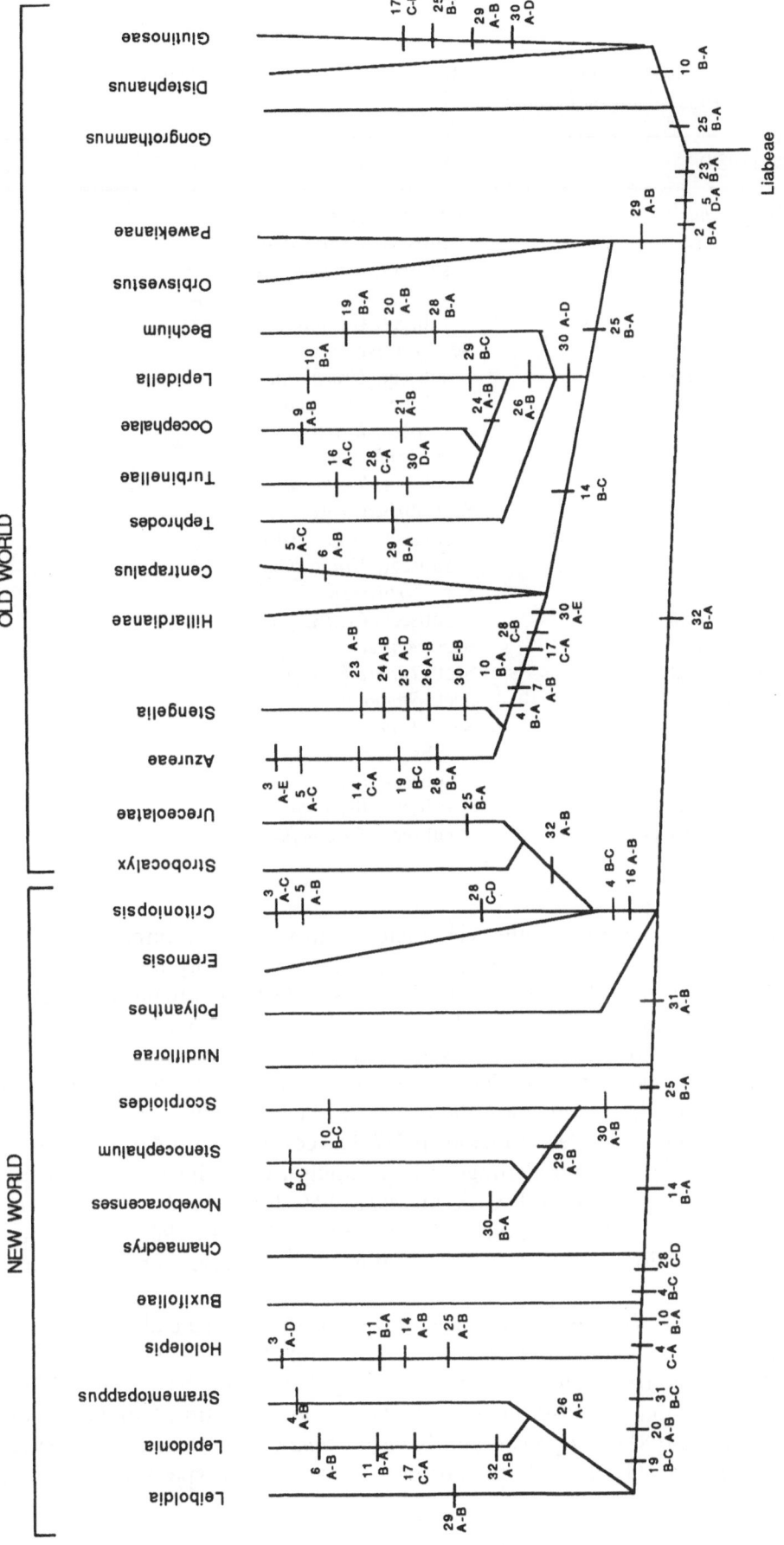

Fig. 2. Cladogram of the sections and subsections of *Vernonia*. Numbers correspond to characters and character states in Table 3; character state changes indicated by letters. Tree length 91, CI=0.52

Table 4. Comparison of JEFFREY's (1988) and JONES' (1981) classifications of Old World sections and subsections of *Vernonia*

JEFFREY (1988)	JONES (1981)
Sect. *Distephanus*	Sect. *Orbisvestus*
subsect. *Distephanus*	subsect. *Distephanus*
	subsect. *Gongrothamnus*
subsect. *Glutinosae*	Sect. *Tephrodes*
	subsect. *Glutinosae*
Sect. *Gymnanthemum*	Sect. *Orbisvestus*
subsect. *Gymnanthemum*	subsect. *Urceolatae*
subsect. *Monosis*	subsect. *Strobocalyx*
	subsect. *Pawekianae*
subsect. *Cyanthillium*	Sect. *Tephrodes*
	subsect. *Tephrodes*
	Sect. *Orbisvestus*
	subsect. *Orbisvestus*
	subsect. *Hilliardianae*
Sect. *Centrapalus*	Sect. *Orbisvestus*
subsect. *Centrapalus*	subsect. *Centrapalus*
subsect. *Linzia*	Sect. *Azureae*
subsect. *Acilepis*	Sect. *Tephrodes* p.p.
Sect. *Baccharoides*	Sect. *Stengelia*
Sect. *Crystallopollen*	Sect. *Tephrodes*
	subsect. *Oocephalae*
	subsect. *Lepidella*
− − − (not treated)	subsect. *Bechium*
− − − (not treated)	subsect. *Turbinellae*

shows the frequent changes in these characters. Chromosome number (31), chemistry (32), corolla color (5), and phyllotaxy (1) are synapomorphies, but these do not operate independently of the other characters. Other data are needed to clarify the relationships among the clades. The clades shown here appeared consistently in all analyses, regardless of variations in outgroup coding, or taxa. For example, we used the *Vernonieae* genera *Piptocarpha* and *Stokesia* as outgroups instead of the *Liabeae*. The clades remained the same, as they did when selected *Liabeae* characters were changed (i.e., to include pale). The constant nature of these groups supports classically recognized subgeneric groupings, e.g., GLEASON 1906, 1922, 1923; EKMAN 1914; BLAKE 1931; CUATRECASAS 1956; JONES 1979 c, 1981; JEFFREY 1988; KEELEY 1978, 1982; TURNER 1981; ROBINSON & KAHN 1986; ROBINSON & FUNK 1987), but does not provide strong support for generic segregates in most cases.

Gongrothamnus, Distephanus, and *Glutinosae*, are distinct from all other sections and subsections of *Vernonia*. These taxa share with the outgroup, the *Liabeae,* yellow flowers and trinervate leaves, but differ in pollen, anther base, and by having alternate rather than opposite leaves. In the synoptic treatment of Old World *Vernonia* (JONES 1981) *Distephanus* and *Gongrothamnus* were treated as subsections of sect. *Orbisvestus* (Table 4). Both subsections share pollen type A, considered to be the primitive type in the genus (KEELEY & JONES 1977 a, 1979). JONES (1981) recognized subsect. *Glutinosae* as distinct, placing it in sect. *Tephrodes* on the basis of its highly derived type E pollen, shared by other members of *Tephrodes*. ROBINSON & KAHN (1986) synonymized subsectt. *Distephanus* and *Gongrothamnus*, as recognized by JONES (1981), raising *Distephanus* to generic level on the basis of the distinctive yellow flower color, trinervate leaves, tailed anthers, stylar node, and pollen type. Although they treated some of the same taxa as recognized by JONES in his subsect. *Glutinosae*, and noted the pollen variation, they did not place this subsection into synonymy. JEFFREY (1987, and pers. comm.) concurred with ROBINSON & KAHN (1986) that *Distephanus* and *Gongrothamnus* should be treated together, but JEFFREY (1988) retained them under the genus *Vernonia* (Table 4). JEFFREY (1988) also felt that the obvious pollen difference noted by JONES (1981) should be recognized and so identified *Vernonia* sect. *Distephanus* subsect. *Distephanus* (incl. *Gongrothamnus*) and subsect. *Glutinosae* (Table 4).

While these yellow-flowered taxa warrant recognition, it is not yet apparent if they should be separated from *Vernonia*. WILD (1978) described a number of African mainland *Vernonia* spp. with trinervate leaves, but purple rather than yellow flowers, and type A pollen. Morphologically they appear to be very similar to the yellow-flowered taxa and share the scandent habit, which is rare among other taxa. ROBINSON & KAHN (1986) felt that while these African purple-flowered taxa were outside the apparent center of distribution in Madagascar, they were nonetheless members of *Distephanus*. JEFFREY (1988) feels that these taxa are more closely allied to other species in *Vernonia* and recognized them as *Vernonia* sect. *Gymnanthemum*. This treatment removed part of JONES' (1981) subsect. *Urceolata* (Table 4). ROBINSON & KAHN (1986) recognized *Gymnanthemum coloratum* at the generic level, but did not relate it to *Distephanus*. It is clear that further study is needed to elucidate the relationships of the yellow and purple-flowered, trinervate species. For the time being, separate generic status for the yellow-flowered species (Table 1) may provide taxonomic convenience, but requires further support.

The problem with flower color, and trinervate leaves is also compounded by the presence of trinervy in at least one New World taxon, *Vernonia pedunculata* DC. of sect. *Hololepis*. This Brazilian taxon has leafy trinervate bracts subtending the involucre. Although the cauline leaves are pinnately-veined, it is apparent that genes for trinervy exist, if only expressed near the inflorescence. If trinervy is taken as basal, then sect. *Hololepis* would appear to be closer to the Old World trinervate taxa than to the New World taxa with which it is now associated. This taxon has the typical purple flower color of most Vernonias. Unfortunately, few specimens of this taxon are available and nothing is known of its chromosome number or chemistry. It would be extremely desirable to obtain more information on this little studied taxon. It may be that divergence between new and Old World lineages is very ancient indeed, and perhaps involved a reticulated split which left at least some connections between taxa now geographically remote and morphologically distant.

At the far end from *Gongrothamnus* and *Distephanus/Glutinosae* is the clearly marked clade of *Leiboldia* (JONES 1979 b), *Lepidonia* (TURNER 1981), *Stramentopappus* (ROBINSON & FUNK 1987). These section/genera share a simple few-headed terminal capitulescence, generally of many flowers per head, many series of foliaceous or subfoliaceous phyllaries, glabrous 5-angled achenes, deciduous pappus and a chromosome number of x = 19. The taxa in these groups appear to form a well supported monophyletic group, although there is disagreement on their taxonomic status. TURNER (1981) pointed out the distinctive nature of *Lepidonia*, and rearranged the taxa treated by JONES (1979 b) as sect. *Leiboldia*. TURNER (1981) regarded *Lepidonia* as primitive and possibly related to African or Old World taxa, on the basis of chromosome number (n = 19), which could be derived by aneuploidy or amphidiploidy from x = 9 or 10, the common number in Old World taxa, and chemistry. *Vernonia jonesii* B. TURNER of sect. *Lepidonia* contains non-glaucolide germacranolides of the type typically found in Old World taxa (GERSHENZON & al. 1984) while *V. pooleae* B. TURNER (*Stramentopappus pooleae* sensu ROBINSON & FUNK 1987) has glaucolide B, typical of New World elements. ROBINSON & FUNK (1987) feel that these taxa are all autochthonous New World elements and that *Lepidonia* is derived. They also felt it was cladistically acceptable to retain *V. pooleae* within *Lepidonia*, but preferred to give it generic rank. Removing the *Lepidonia*/*Leiboldia*/*Stramentopappus* clade does leave a paraphyletic assemblage in *Vernonia* as recognized by ROBINSON & FUNK (1987). They did not include the monotypic sect. *Hololepis* in their analysis. This is seemingly the next closest taxonomic unit and one whose affinities, as discussed above under trinervy, need further clarification. It is not yet possible to say with certainty whether the *Leiboldia* group, or part of it, are related to Old World elements, but this should be reexamined as additional data become available.

The sect. *Hololepis* has been recognized as distinct within *Vernonia* since its original description by DE CANDOLLE (1836). It is unusual, not only because of its trinervate bracts, a feature incorrectly illustrated in "Flora Brasiliensis" (BAKER 1873), but also because it is paleate, a feature also not noted in early work. The distinct position of this taxon is confirmed here. It appears near *Lepidonia*, which includes the other paleate taxon *Vernonia paleata* BLAKE (*Lepidonia paleata* sensu ROBINSON & FUNK 1987).

Between the well-marked clades of the *Distephanus* group on the one end and the *Leiboldia* group on the other, are the two clades which include most species of *Vernonia*. The Old and New World taxa are largely distinct from one another, although there are areas where the relationships are less clear (Figs. 1 and 2). This is in agreement with the treatment of JONES (1979c, 1981) (Table 2) who erected two distinct subgenera for these two elements based on pollen, chemistry, chromosome number and morphology. JEFFREY (1988) also noted the prominent differences in pollen, chemistry and chromosome number, supporting the idea of distinct generic status for the New and Old World subgenera if they could be shown to be monophyletic sister groups. In his recent work, JEFFREY (1988) revised the subgeneric categories proposed by JONES (1981) based on his work with African taxa for the "Flora of Tropical East Africa" (Table 4). A number of taxa were maintained separate following traditional logic, i.e., *Adesia* which share features with *Vernonia*, but new relationships were also recognized.

Our analysis (Figs. 1 and 2) provides support for a number of the rearrangements suggested by JEFFREY (1988), but also documents the need for further study. Sect. *Crystallopollen* reflects his revision, placing JONES' subsectt. *Oocephalae* and *Lepidella* next to one another in the same clade. The other subsectt. *Turbinellae* and *Bechium*, were not treated in JEFFREY's revision but also appear with this group. Their position needs confirmation. Members of his sect. *Gymnanthemum* include

JONES' subsectt. *Orbisvestus* and *Pawekianae*, which also group together. Other members of this section are not as closely placed in this analysis (Table 4). JONES' subsectt. *Hilliardianae* and *Tephrodes* separate from each other on pollen type and habit, but *Strobocalyx* and *Urceolatae* group together, albeit at some distance from the larger clade of Old World groups. Members of JEFFREY's sect. *Centrapalus* include JONES' sect. *Azureae* and subsect. *Centrapalus*, an arrangement supported by their presence in the same clade. JEFFREY (1988) commented that the stengelioid taxa, which he places in sect. *Baccharoides* form one of the taxonomically most difficult groups of African *Vernonieae*. Given the great variability of many taxa, and evidence of hybridization, species limits are not clear and much study is needed. The delimitation of *Stengelia* is certain to undergo modification with additional study. In our analysis its position is more a reflection of the difficulty of incorporating wide variation, than it is a definitive statement on position. Generic status may be justified, but would appear to need further support.

The New World subsections proposed by JONES (1979) are largely supported in this analysis, although there is considerable variation within the numerous series of subsect. *Scorpioides* (KEELEY 1982, and unpubl.). ROBINSON (1987) has proposed that subsect. *Stenocephalum* be raised to generic status on the basis of the small number of florets per head, the position of heads in the capitulescence, the stylar base and the uniformity of the carpopodium. Although these characters may be useful at the specific level, and perhaps at higher levels, they did not serve to remove these taxa from the clade containing most New World taxa in our analysis. The chromosome counts of $n = 17$ for *Vernonia megapotamica* (JONES 1979 a) and $n = 12$ for *Vernonia jucunda* (STROTHER 1983), along with the morphological features suggests that further work is needed to confirm the relationship of this group to other New World subsections. Subsectt. *Buxifoliae* and *Chamaedrys* are morphologically distinct from all other New World vernonias with large turbinate involucres of many series of tightly imbricated bracts. They are geographically isolated from each other, the subsect. *Buxifoliae* being confined to the island of Hispaniola in the Caribbean, whereas the subsect. *Chamaedrys* are found in the interior of S. America. There may be a close relationship between these groups, this needs further investigation. It would also appear that the taxonomic status of these may need reevaluation since they stand apart from the clade containing most of the New World taxa.

The subsect. *Eremosis*, including the S. American members which have been recognized as the sect. *Critoniopsis* (CUATRECASAS 1956) or the genus (ROBINSON 1980), are separated from the clade which contains most of the New World subsections (Figs. 1 and 2). Indeed, *Eremosis/Critoniopsis* grouped with subsectt. *Strobocalyx* and *Urceolatae* of the Old World. Both Old and New World taxa in these groups typically have less than 10 flowers per head (like taxa of subsect. *Stenocephalum*) and deciduous phyllaries. These attributes may represent parallel evolution in geographically distinct lines, however, this is not yet clear. Because of the clear count of $n = 10$ for *Vernonia obtusa* (TURNER 1981) of subsect. *Eremosis* and anomalous counts which may represent tetraploids on $x = 10$ for other members of this subsection (JONES 1973) the connection may be more basic. The S. American members have been considered to be closely related to the genus *Piptocarpha* (ROBINSON 1980) which has as base chromosome number of $x = 17$. In the present analysis, and one by KEELEY & DIAZ (unpubl.) there were insufficient differences between the Mexican (*Eremosis*) and S. American (*Critoniopsis*) taxa to warrant generic status for *Critoniopsis* apart from *Eremosis*. ROBINSON (1980) did not consider the Mexican (*Eremosis*) taxa in raising the Andean species to generic level. The position of sect. *Eremosis* outside the clade containing most other New World *Vernonia* subsections suggests a more distant relationship, which may justify recognition at the generic level. Certainly, a base chromosome number of x = 10, and a strong morphological relationship to Old World taxa, indicates that further study is needed to clarify its phyletic position.

The analysis shown here provides support for repeated parallelisms and reversals within a number of morphological characters such as habit, number of florets per head, capitulescence type, stylar trichomes, style base, phyllary type and series, which have been used to distinguish *Vernonia* taxa at all levels. The presence of trinervy in both Old and New World members, the equivocal nature of chromosome counts and the lack of information in many taxa indicate that much work remains to be done, and that splitting of genera from this assemblage is far from a clear solution to the problem. Certain clades appear stable; the *Distephanus* — *Glutinosae* — *Gongrothamnus* group, the Old World vernonias, the New World vernonias, the *Eremosis-Critoniopsis* and *Strobocalyx-Urceolatae* clade, and the New World groups *Hololepis,* and *Leiboldia, Lepidonia* and *Stramentopappus*, but their precise relationships are still much in need of clarification. The phylogenetic relationships indicated by the two cladograms are different. In one, some New World taxa at least, are derived near the base of the Old World lineage, and are still strongly connected to it (Fig. 1). In the other, the New and Old World taxa are largely distinct, except for two African subsections which are derived from the line giving rise to New World taxa (Fig. 2). Data are lacking to resolve these substantive differences, but are required before generic segregations can be made with confidence.

Research on the *Vernonieae* has been supported by NSF grants DEB 79-04757 and BSR 82-07010 to SCK, and by Whittier College. PAUL DELANEY, GORDON HENDLER, ROBERT K. JANSEN, GUY NESOM, FLORENCE NISHIDA, DON R. REYNOLDS, NOELLE SEDOR, ROBERT LAVENBERG and the Sections of Botany, Invertebrate Zoology, and Icthyology, Natural History Museum of Los Angeles, provided support and inspiration during the course of this work.

Appendix 1. Data matrix for cladistic analysis of *Veronia*. Unknown or inapplicable characters. Characters discussed in text.

```
LEIBOLD   AAAAAAAAAABDBAAACACBAAAAAABDBAC?
LEPIDON   AAAAABAAAAABBAAAAACBAAAAABBDAACB
STRAMEN   AAABAAAAAABBBAAACACBAAAAABBDAACA
HOLOLEP   AADAAAAAAAABBBAACABAAAAABAADAA?A
NOVEBOR   AAABAAAAABBCBBAACBBAAAAAAAACBABA
EREMOSI   AAACAAAAABBDABABCBBAAAAABABCAAAA
CRITONI   AACCBAAAABBDABABCBBAAAAABAADAAAA
POLYANT   AAABAAABABBBBBAACBBAAAAABAACAAAA
CHAMEDR   AAABAAAAABBCBAAACBBAAAAAAAACAABA
BUXIFOL   AAACAAAAABBCBAAACBBAAAAAAADAABA
SCORPIO   AAABAAAAACBCBBAACBBAAAAAAAACABBA
STENOCE   AAACAAAAABBBABAACBBAAAAAAAACBBBA
NUDIFLO   AAABAAAAABBCABAACBBAAAAABAACAABA
STROBOC   AAACAAAAABBBBBABCBBAAAAABAACAAAB
GONGROT   ABABDAABABBBABAACBBAAABAAABCAAAB
DISTEPH   ABABDAAAAABCBBAACBBAAABAAAACAAAB
URCEOLA   AAACAAAAABBCABABCBBAAAAAAABCAAAB
AZUREAE   AAEACABBAABDBABAAACAAAAAAABEAB
STENGEL   AAAAAABAAABCBCBAAABAAABBDBABBBAB
TEPHROD   AAABAAAAABBBBBBACBBAAAAAAAACADAB
ORBISVE   AAABAAAAABBBBBAACBBAAAAABABDBAAB
PAWEKIA   AAABAAAAABBCABBACBBAAAAABAACBAAB
HILIARD   AAABAAAAABBCACBACBBAAAAAAABCBAAB
TURBINE   AAABAAABABBCABACCABAAAABABBABAAB
CENTAFRO  AAABCBAAABBCACBACBBAAAAAAABCBDAB
LEPIDELL  AAABAAAAAABBABBACBBAAAAAABBCCDAB
OOCEPHAL  AAABAAAABBBCABAACBBABAABABACBDAB
GLUTINOS  ABABDAAAAABBBBBABABAAABABAACBDAB
BECHIUM   AAABAAAAABBCBBAACAABAAAAAAACBDAB
CACOSMIA  BBACDCBABBBCABABCCDCCCBABABCA?A?
LIABUM    BBABDAABBBBBBAAACAAABABABABCA?A?
```

References

ARISTEGUIETA, L., 1963: El genero *Oliganthes* de Madagascar y su equivalente americano *Pollalesta*. — Soc. Venez. Cienc. Nat. **23**: 255—289.

AUGIER, J., DUMERAC, M. L., 1951: La phylogénie des composées. — Rev. Sci. **3311**: 167—182.

BAKER, J. G., 1872: *Compositae* I. *Vernoniaceae*. — In MARTIUS, C. F. P. VON (Ed.): Flora Brasiliensis **6**, pp. 1—179, pl. 1—50.

BENTHAM, G., 1873: Notes on the classification, history, and geographical distribution of the *Compositae*. — J. Linn. Soc. Bot. **13**: 335—577.

— HOOKER, J. D., 1873: Genera Plantarum 2 (1). London.

BLAKE, S. F., 1936: *Lepidonia*, a new genus of *Vernonieae*, with a nomenclatorial note on the name *Leiboldia*. — J. Washington Acad. Sci. **26**: 452—460.

BOHLMANN, F., JAKUPOVIC, J., 1990: Progress in the chemistry of the *Vernonieae*. — Pl. Syst. Evol. [Suppl. 4]: 3—43.

BOLICK, M. R., 1978: A light and electron microscope study of pollen of the *Vernonieae* (*Compositae*). — Ph.D. Thesis, University of Texas, Austin.

BREMER, K., 1987: Tribal interrelationships of the *Asteraceae*. — Cladistics **3**: 210—253.

BUSEY, P., 1975: Flora of Panama: *Elephantopodinae*. — Ann. Missouri Bot. Gard. **62**: 873—879.

CARLQUIST, S., 1976: Tribal interrelationships and phylogeny of the *Asteraceae*. — Aliso **8**: 465—492.

CASSINI, H., 1825: Vernoniees. − In CUVIER, G. L., (Ed.): Dictionnaire des sciences na-
turelles **25**, pp. 338−345. Reprinted in KING, R. M., DAWSON, H. L., (Eds.): CASSINI
on *Compositae*. − New York: Oriole Editions.

CLONTS, J. A., 1972: A revision of *Elephantopus* and *Pseudoelephantopus*. − Ph.D.Thesis,
Mississippi State University, Starkville, MS.

COILE, N., JONES, S., 1981: Lychnophora (Compositae: Vernonieae), a genus endemic to
the Brazilian Planalto. − Brittonia **33**: 528−542.

COLEMAN, J. R., 1968: Chromosome numbers in some Brazilian *Compositae*. − Rhodora
70: 228−240.

CRONQUIST, A., 1955: Phylogeny and taxonomy of the *Compositae*. − Am. Midl. Naturalist
53: 478−511.

− 1977: The *Compositae* revisited. − Brittonia **29**: 137−153.

CUATRECASAS, J., 1956: Neue *Vernonia*-Arten und Synopsis der andinen Arten der Sektion
Critoniopsis. − Bot. Jahrb. Syst. **77**: 52−84.

DE CANDOLLE, A. P., 1836: *Composiae*. − Prodromus Systematics **5**: 4−706.

DILLON, M., TURNER, B. L., 1982: Chromosome numbers of Peruvian *Compositae*. −
Rhodora **84**: 131−137.

EKMAN, E. L., 1914: West Indian *Vernonieae*. − Ark. Bot. **13**: 1−106.

FAUST, W. Z., JONES, S. B., Jr., 1973: The systematic value of trichome complements in a
North American group of *Vernonia* (*Compositae*). − Rhodora **75**: 517−528.

FUNK, V. A., 1982: The systematics of *Montana* (*Asteraceae, Helintheae*). − Mem. New
York Bot. Gard. **36**: 1−133.

− 1985: Phylogenetic patterns and hybridization. Ann. Missouri Bot. Gard. **72**: 681−715.

GERSHENZON, J., PFEIL, R. M., LIU, Y. L., MABRY, T. J., TURNER, B. L., 1984: Sesquiterpene
lactones from two newly-described species of *Vernonia*: *V. jonesii* and *V. pooleae*. −
Phytochemistry **22**: 777−780.

GLEASON, H. A., 1906: A revision of the North American *Vernonieae*. − Bull. New York
Bot. Gard. **4**: 144−243.

− 1922: *Vernonieae*. − North Am. Fl. **33**: 52−95.

− 1923: Evolution and geographical distribution of the genus *Vernonia* in North America.
− Am. J. Bot. **10**: 187−202.

HUMBERT, H., 1960: Composées: Vernoniées. − Flore de Madagascar **189**: 1−199.

JANSEN, R. K., 1985: The systematics of *Acmella* (*Asteraceae: Heliantheae*). − Systematic
Botany Monographs **8**.

− STUESSY, T. F., DIAZ-PIEDRAHITA, S., FUNK, V. A., 1984: Recuentos chromosomicos
en *Compositae* de Colombia. − Caldasia **14**: 7−20.

− SMITH, E. B., CRAWFORD, D. A., 1987: A cladistic study of North American *Coreopsis*
(*Asteraceae: Heliantheae*). − Pl. Syst. Evol. **157**: 73−84.

JEFFREY, C., 1988: The *Vernonieae* in east tropical Africa. Notes on the *Compositae*:
V. − Kew Bull. **43**: 195−277.

JONES, S. B., Jr., 1976: Cytogenetics and affinities of *Vernonia* (*Compositae*) from the
Mexican highlands and eastern North America. − Evolution **30**: 544−462.

− 1977: *Vernonieae* systematic review. − In HEYWOOD, V. H., HARBORNE, J. B., TURNER,
B. L., (Eds.): The biology and chemistry of the *Compositae* 1, pp. 503−521. − London:
Academic Press.

− 1979 a: Chromosome numbers of *Vernonieae* (*Compositae*). − Bull. Torrey Bot. Club
106: 79−84.

− 1979 b: Taxonomic revision of *Vernonia* sect. *Leiboldia* (*Compositae: Vernonieae*). −
Castanea **44**: 229−237.

− 1979 c: Synopsis and pollen morphology of *Vernonia* (*Compositae: Vernonieae*) in the
New World. − Rhodora **81**: 425−447.

− 1981: Synoptic classification and pollen morphology of *Vernonia* (*Compositae: Vernonieae*) in the Old World. − Rhodora **83**: 59−75.

KEELEY, S. C., 1978: A revision of the West Indian vernonias (*Compositae*). — J. Arnold Arbor. **59**: 360 – 413.

— 1982: Morphological variation and the problem of species recognition in the neotropical taxon *Vernonia arborescens* (*Compositae*). — Syst. Bot. **7**: 71 – 84.

— JONES, S. B., Jr., 1977 a: Taxonomic implications of external pollen morphology to *Vernonia* (*Compositae*) in the West Indies. — Am. J. Bot. **64**: 576 – 584.

— — 1977 b: *Vernonia* (*Compositae*) in the Bahamas re-visited. — Rhodora **79**: 147 – 159.

— — 1979: Distribution of pollen types in *Vernonia* (*Vernonieae: Compositae*). — Syst. Bot. **4**: 195 – 202.

KING, B. L., JONES, S. B., Jr., 1982: Chemosystematics of *Vernonia* series *Flexuosae* (*Vernonieae: Compositae*). — Bull. Torrey Bot. Club **109**: 279 – 286.

KINGHAM, D. L., 1976: A study of the pollen morphology of tropical African and certain other *Vernonieae* (*Compositae*). — Kew Bull. **31**: 9 – 26.

KIRKMAN, L. K., 1981: Taxonomic revision of *Centratherum* and *Phyllocephalum* (*Compositae: Vernonieae*). — Rhodora **83**: 1 – 24.

KITAMURA, S., 1968: Taxonomic studies in the genus *Camchaya*. — Acta Phytot. Geobot. **23**: 71 – 73.

MABRY, T., ABDEL-BASET, PANDOLINA, F. JONES, S. B., Jr., 1975: Systematic implications of flavonoids and sesquiterpene lactones in species of *Vernonia*. — Biochem. Syst. Evol. **2**: 185 – 192.

MACLEISH, N., 1984 a: Revision of *Eremanthus* (*Compositae: Vernonieae*). — Ph.D. thesis, University of Georgia, Athens, GA.

— 1984 b: *Argyrovernonia* and *Paralychnophora*: new names in the tribe *Vernonieae*. — Taxon **33**: 106 – 107.

— 1985 a: Revision of *Glaziovianthus* (*Compositae: Vernonieae*). — Syst. Bot. **10**: 347 – 352.

— 1985 b: Revision of *Chresta* and *Pycnocephalum* (*Compositae: Vernonieae*). — Syst. Bot. **10**: 459 – 470.

— 1987: Revision of *Eremanthus* (*Compositae: Vernonieae*). — Ann. Missouri Bot. Gard. **74**: 265 – 290.

MEACHAM, C., 1984: The role of hypothesized direction of characters in the estimation of evolutionary history. — Taxon **33**: 26 – 38.

NORDENSTAM, B., 1977: *Senecioneae* and *Liabeae* — systematic review. — In HEYWOOD, V. H., HARBORNE, J. B., TURNER, B. L., (Eds.): The biology and chemistry of the *Compositae* 2, pp. 799 – 830. — London: Academic Press.

POPE, G., 1983: Cypselas and trichomes as a source of taxonomic characters in the erlangeoid genera. — Kirkia **12**: 203 – 231.

RABAKNANDRIANINA, E., CARR, G. D., 1987: Chromosome numbers of Madagascar plants. — Ann. Missouri Bot. Gad. **74**: 123 – 125.

ROBINSON, H., 1977: An analysis of the characters and relationships of the tribes *Eupatorieae* and *Vernonieae* (*Asteraceae*). — Syst. Bot. **2**: 199 – 208.

— 1979: Two new genera of *Vernonieae* (*Asteraceae*) from Brazil, *Heterocypsela* and *Pseudostifftia*. — Phytologia **44**: 442 – 449.

— 1980: Re-establishment of the genus *Critoniopsis*. — Phytologia **46**: 437 – 442.

— 1981: *Episcothamnus* and *Bishopalea*, two new genera of the *Vernonieae* (*Asteraceae*) from Brazil, and the resurrection of *Sipolisia*. — Phytologia **48**: 209 – 220.

— 1983: A generic review of the tribe *Liabeae* (*Asteraceae*). — Smithsonian Contrib. Bot. **54**: 1 – 69.

— 1987: Studies of the *Lepidaploa* complex (*Vernonieae: Asteraceae*). 1. The genus *Stenocephalum* SCH. BIP. — Proc. Biol. Soc. Washington **100**: 578 – 583.

— BRETTEL, R. D., 1973: Tribal revisions in the *Asteraceae*. 3. A new tribe, *Liabeae*. — Phytologia **25**: 404 – 407.

— FUNK, V., 1987: A phylogenetic analysis of *Leiboldia*, *Lepidonia*, and a new genus *Stramentopappus* (*Vernonieae: Asteraceae*). — Bot. Jahrb. Syst. **108**: 213 – 228.

– Kahn, B., 1986: Trinervate leaves, yellow flowers, tailed anthers, and pollen variation in *Distephanus* Cassini (*Vernonieae: Asteraceae*). – Proc. Biol. Soc. Washington **99**: 493–501.

– King, R. M., 1979: *Mattfeldanthus mutisioides* gen. et spec. nov. (*Asteraceae: Vernonieae*) from Bahia, Brazil. – Willdenowia **9**: 9–12.

– Bohlmann, F., King, R. M., 1980: Chemosystematic notes on the *Asteraceae*. 3: Natural subdivisions of the *Vernonieae*. – Phytologia **46**: 421–436.

Seaman, F. C., 1982: Sesquiterpene lactones as taxonomic characters in the *Asteraceae*. – Bot. Rev. **48**: 121–595.

Skvarla, J. J., Turner, B. L., Tomb, A. S., 1977: Pollen morphology in the *Compositae* and in morphologically related families. – In Heywood, V. H., Harborne, J. B., Turner, B. L., (Eds.): The biology and chemistry of the *Compositae*. – London: Academic Press.

Smith, C. E., 1971: Observations on the stengelioid species of *Vernonia*. – U.S.D.A. Agric. Handbook **396**.

Smith, G., 1982: Taxonomic consideration of *Piptocarpha* (*Compositae: Vernonieae*) and new taxa in Brazil. – Brittonia **34**: 210–218.

– 1984: A taxonomic revision of *Piptocarpha* (*Compositae: Vernonieae*). – Ph.D. Thesis, University of Georgia, Athens, GA.

Strother, J. L., 1983: More chromosome studies in *Compositae*. – Am. J. Bot. **70**: 1217–1224.

Stutts, J. G., 1981: Taxonomic revision of *Pollalesta* H. B. K. (*Compositae: Vernonieae*). – Rhodora **83**: 385–419.

– Muir, M. A., 1981: Taxonomic revision of *Piptocoma* Cass. (*Compositae: Vernonieae*). – Rhodora **83**: 77–86.

Sundberg, S., Cowan, C. P., Turner, B. L., 1986: Chromosome counts of Latin American *Compositae*. – Am. J. Bot. **73**: 33–38.

Swofford, D. L., 1985: PAUP – phylogenetic analysis using parsimony, version 2.4.0. – Champaign: Illinois Natural History Survey.

Turner, B. L., 1977: Summary of the biology of the *Compositae*. – In Heywood, V. H., Harborne, J. B., Turner, B. L., (Eds.): The biology and chemistry of the *Compositae* 2, pp. 1105–1118. – London: Academic Press.

– 1981: New species and combinations in *Vernonia* sections *Leiboldia* and *Lepidonia* (*Asteraceae*), with a revisional conspectus of the groups. – Brittonia **33**: 401–412.

– Powell, A. M., Cuatrecasas, J., 1967: Chromosome numbers in *Compositae* 11. Peruvian species. – Ann. Missouri Bot. Gard. **54**: 172–177.

Wagenitz, G., 1976: Systematics and phylogeny of the *Compositae* (*Asteraceae*). – Pl. Syst. Evol. **157**: 29–46.

Watrous, L., Wheeler, Q., 1981: The outgroup method of character analysis. – Syst. Zool. **30**: 1–11.

Wild, H., 1978: The *Compositae* of the Flora Zambesiaca area 8. *Vernonieae* (*Vernonia*). – Kirkia **11**: 31–127.

Wild, H., Pope, G. V., 1977: The *Compositae* of the flora Zambesiaca area, 7. – *Vernonieae* (excluding *Vernonia* Schreb.). – Kirkia **10**: 309–338.

Willis, J. C. (revised by Shaw, H. K. A.) 1985: A dictionary of flowering plants and ferns, 8th edn. – New York: Cambridge University Press.

Addresses of the authors: Sterling C. Keeley, Department of Biology, Whittier College, Whittier, CA 90608, U.S.A. – Billie L. Turner, Department of Botany, University of Texas, Austin, TX 74713, U.S.A.

Pl. Syst. Evol. [Suppl. 4], 67 – 75 (1990)

Chemistry of the *Heliantheae* (*Compositae*)

F. BOHLMANN

Received December 4, 1987

Key words: Angiosperms, *Compositae, Heliantheae.* – Phytochemistry, sesquiterpene lactones, polyacetylenes.

Abstract: In connection with the separation of the *Heliantheae* into subtribes, the relevant chemistry is discussed briefly. The different types of sesquiterpene lactones and polyacetylenes are the most characteristic features. However, in several groups other constituents are also useful. Still, not in all cases chemistry allows for a clear differentiation.

The tribe *Heliantheae* with c. 3000 species is separated into 260 genera, and these are placed in the subfam. *Asteroideae* close to *Inuleae* and *Eupatorieae.* A revision of the subtribal limits led to 35 subtribes (ROBINSON 1981). In part these subtribes can be characterized by their main chemical constituents (Table 1). Most widespread in the tribe are polyacetylenes (Fig. 1), sesquiterpene lactones (Fig. 2) and diterpenes, mainly kaurane but also labdane derivatives (Fig. 3). Furthermore, prenylated p-hydroxyacetophenones, phenylpropanes and some special compounds are accumulated in some subtribes (Fig. 4). In the case of the sesquiterpene lactones, germacran-12,6-olides with 8 β-oxygen function are most wide-spread, often as heliangolides or furoheliangolides. However, these lactones are mostly concentrated in a few subtribes. The general position of *Heliantheae* (and *Vernonieae*) within the *Compositae* as visualized by their chemistry is shown in Fig. 5.

If we follow the scheme of ROBINSON (Table 1) the first subtribe is *Ambrosiinae* which is characterized by pseudoguaianolides and in part by seco guaianolides. In the *Espeletiinae* no sesquiterpene lactones have been reported but in all genera kaurene derivatives are accumulated, while the *Melampodiinae* are characterized by 1 (10) Z-germacranolides, the melampolides. These are absent in the next subtribe *Polymninae* which supports the separation of *Smallanthus* and *Polymnia.* In the *Milleriinae* and also in the *Enhydrinae*, melampolides again predominate. The separation of *Montanoa* as a new subtribe is supported by the occurrence of rare cis-6,12-germacranolides. In the subtribe *Rudbeckiinae* very different sesquiterpene lactones are present; typical polyacetylenes derived from the pentaynene may be characteristic for this group.

Many genera have been placed in the subtribe *Ecliptinae* which chemically do not present an uniform picture. Parts of these genera have very characteristic sesquiterpene lactones, eudesmanolides and elemanolides with an unusual 10-α-methyl group. They are present in *Aspilia, Steiractinia, Wedelia, Zexmenia,* and

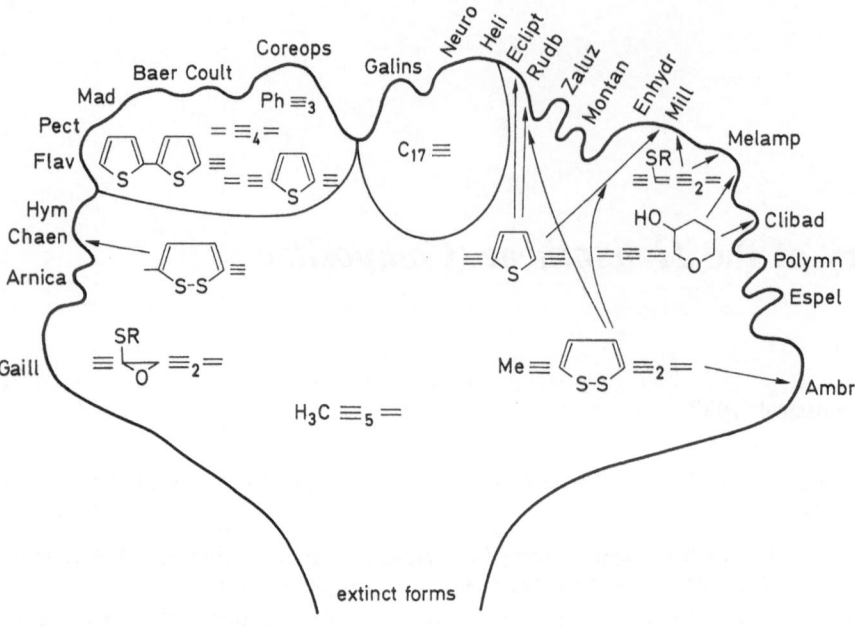

Fig. 1. Distribution of polyacetylenes among subtribes of *Heliantheae* (abbreviations cf. Table 1) and the genus *Arnica*

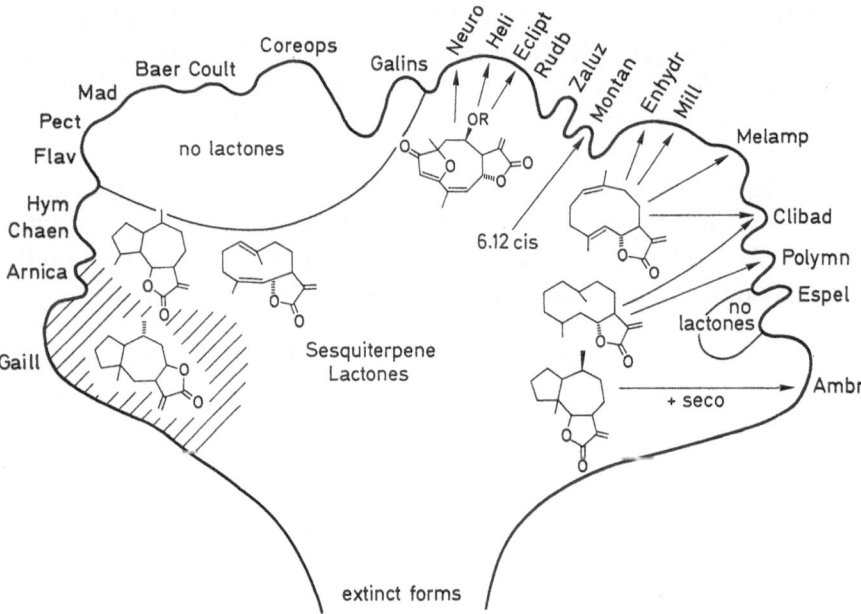

Fig. 2. Distribution of sesquiterpene lactones among subtribes of *Heliantheae* and the genus *Arnica*

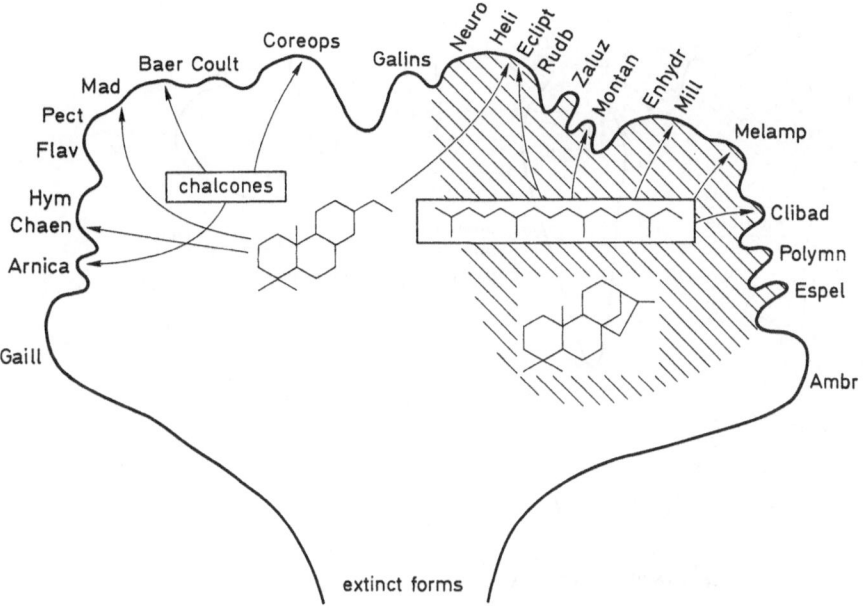

Fig. 3. Distribution of diterpene among subtribes of *Heliantheae* and the genus *Arnica*

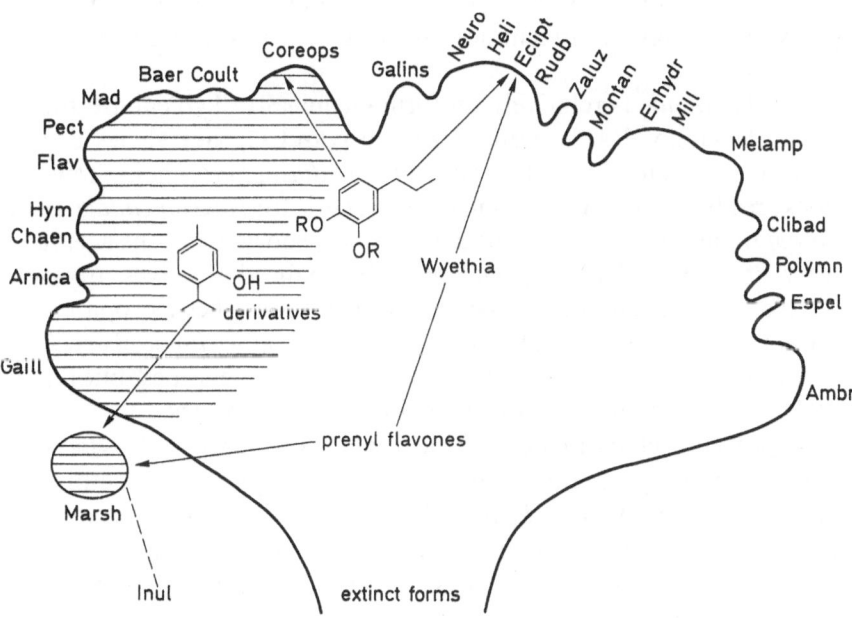

Fig. 4. Distribution of thymol and phenyl propane derivatives among subtribes of *Heliantheae* (including the genus *Arnica*), and their possible links with *Inuleae* (*Inul*)

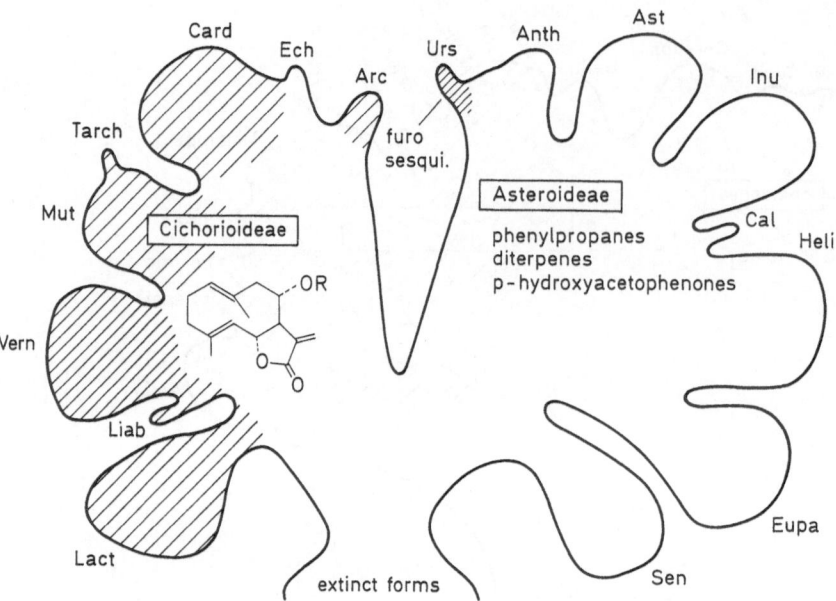

Fig. 5. Position of *Heliantheae* within the *Compositae*

Zinnia. Highly unsaturated amides are reported from *Acmella, Echinacea, Heliopsis, Salmea, Sanvitalia,* and *Spilanthes.* Furthermore, thiophene acetylenes, kaurene derivatives, prenylated p-oxyacetophenones and different types of sesquiterpene lactones are wide-spread.

The subtribe *Helianthinae* is relatively uniform chemically. Typical are the accumulation of furoheliangolides, kaurene derivatives and C_{17}-acetylenes derived from dehydrofalcarinone. These are also typical for *Galinsoginae* where, however, furoheliangolides are absent. These are again charactersitic for *Neurolaeninae.* From the subtribe *Coreopsidinae* no sesquiterpene lactones are reported, but the acetylenic chemistry is very uniform and also special phenylpropane derivatives are wide-spread. *Fitchia,* placed in a separate subtribe, also has these phenylpropanes but none of the characteristic acetylenes. The next subtribes, *Coulterellinae, Pectidinae, Flaveriinae, Jaumeriinae* and, *Madiinae,* all lack sesquiterpene lactones, but typical thiophene acetylenes are present in most of the investigated species.

In the *Hymenopappinae* guaianolides are most wide-spread while the *Baeriinae* again can be characterized by the thiophene acetylenes. Sesquiterpene lactones, in part unique ones, are reported from *Eriophyllum* only. The genera placed in the *Chaenactidinae* mostly contain sesquiterpene lactones, but again there are also thiophene acetylenes and in two genera rare diterpenes of the rosane type. Pseudoguaianolides are isolated from *Arnica* only. Accordingly, this genus is standing somewhat apart and may be better placed in the *Gaillardiinae* where these lactones are characteristic. The chemistry of *Marshallia,* the only genus of the subtribe *Marshalliinae,* does not show clear relationships to other subtribes. In addition to thymol derivatives, prenylated flavanoids are characteristic. These compounds are rare in the *Compositae.* In the tribe *Heliantheae* there is only one report of these flavanoids from *Wyethia* and *Flourensia* (placed in the subtribe *Ecliptinae*) and

Table 1. Distribution of sesquiterpene lactones and some other main constituents among genera of *Heliantheae* subtribes (abbreviations in brackets); GE Germacranolide, Gu guaianolide, PGu pseudoguaianolide, Eu eudesmanolide, S seco, ER eremophilanolide, ME melampolide, EL elemanolide, $\equiv_5 =$ tridecapentaynene, Th. $\equiv$ thiophene acetylenes and SMe $\equiv$ methyl thioethers derived from $\equiv_5$; numbers in brackets indicate number of occurrences; *n* Number of species studied; original data from SEAMAN 1982, own results and from lit.

Heliantheae	n	Main constituents
1. Subtribe *Ambrosiinae* (Ambr)		
Ambrosia	26	PGu (20), SPGu (19), 8.12 GE (6), Gu (3), Eu (3)
Dicoria	1	SGu (1)
Hymenoclea	2	PGu, SPGu
Iva	15	PGu (3), SGu (2), 8.12 Eu (7), SEu (1), IVAX (1), 8.12 GE (1), Gu (2)
Parthenice	1	SGu (1)
Parthenium	21	PGu (20), SPGu (1), SGu (3)
Xanthium	15	SGu (15), Gu (1), ER (2), pren. flavones
2. *Espeletiinae* (Espel)		
Carramboa	2	− kauranes
Coespeletia	5	− kauranes
Espeletia	8	− kauranes
Espeletiopsis	5	− kauranes
Libanothamnus	6	− kauranes
Morithamnus	1	− kauranes
Ruilopezia	12	− kauranes
Tamania	1	− kauranes
3. *Melampodiinae* (Melamp)		
Acanthospermum	3	ME (3)
Ichthyothere	6	ME (2)
Melampodium	18	ME (14), Gu (1), GE (2)
Smallanthus	7	Me (7)
4. *Polymniinae* (Polymn)		
Polymnia	2	GE (1), Eu (1)
5. *Milleriinae* (Mill)		
Guizotia	1	−, $\equiv_5 =$ der.
Milleria	1	ME (17)
Sigesbeckia	3	ME (1)
6. *Desmanthodiinae*	−	
Desmanthodium	3	−, labdane
7. *Clibadiinae* (Clibad)		
Clibadium	8	ME (1), GE (1), (18 spp. flavones), special $\equiv$
Riencourtia	3	−
8. *Pinillosinae*	−	
9. *Guardiolinae*	−	
10. *Enhydrinae* (Enhydr)		
Enhydra	2	ME (2), GE (1)
11. *Montanoinae* (Montan)		
Montanoa	15	6.12 cis GE (12), EL (1), 6.12 cis Eu (1), Gu (5), GE (4)

Table 1 continued

Heliantheae	n	Main constituents
12. *Rudbeckiinae* (Rudb)		
Ratibida	2	SGu (2)
Rudbeckia	3	Eu (1), Gu (1), ER (1), PGu (17), (12 spp. acetylenes der. from Me $\equiv_5 =$)
13. *Zaluzaniinae* (Zaluz)	10	Gu (6), GE (2), Eu (2)
14. *Heptanthinae*	–	
15. *Ecliptinae* (Eclipt)		
Acmella	1	–, amides
Actinomeris	3	–
Aspilia	16	–, kauranes, 8.12 Eu (10 α) (2)
Balsamorhiza	2	6.12 Eu (10), Gu (1)
Baltimora	1	8.12 Eu (1), 6.12 GE (1)
Berlandiera	3	Gu (3)
Blainvillea	3	6.12 GE (2), ME (2)
Borrichia	1	$\equiv_5 =$
Calyptocarpus	1	–
Delilia	1	$\equiv_5 =$
Dimerostemma	4	6.12 Eu (4)
Dugesia	1	ER (1)
Echinacea	3	–, amides
Eclipta	3	–, Th. $\equiv$
Elvira	2	–, $\equiv_5 =$
Encelia	9	8.12 Eu (1), oxyacetoph., $\equiv_5 =$
Enceliopsis	3	– oxyacetoph.
Engelmannia	1	–, $\equiv_5 =$
Flourensia	8	8.12 Eu (1), oxyacetoph., pren Flavones
Geraea	2	– oxyacetoph.
Helianthella	3	– oxyacetoph.
Heliopsis	5	– amides
Iostephane	1	– kaurane
Kingianthus	1	PGu (1)
Lasianthaea	1	–
Leptocarpha	1	6.12 GE (2)
Lindheimera	1	–, $\equiv_5 =$, cycloartanes
Melanthera	2	–, $\equiv_5 =$
Monactis	2	PGu (1), Ca (1)
Oyedaea	6	–, kauranes, $\equiv_5 =$
Perymenium	7	6.12 GE (3), kauranes, 8.12 Eu (1), 6.12 ME (2), 5.12 Eu (1)
Podachaenium	1	Gu (1)
Podanthus	2	6.12 GE (2), He (1)
Rensonia	1	–
Salmea	1	6.12 GE (1), amides
Sanvitalia	2	–
Silphium	6	–, kauranes, labdanes, $\equiv_5 =$
Spilanthes	8	8.12 Eu (1), amides
Steiractinia	2	6.12 Eu (10 α) (1), kauranes, $\equiv_5 =$
Synedrella	2	6.12 GE (1), $\equiv_5 =$

Table 1 continued

Heliantheae	n	Main constituents
Verbesina	41	6.12 EL (3), Eu (1), eudesmane esters
Wedelia	20	8.12 Eu (10 α) (6), kauranes, $\equiv_5=$
Wulffia	1	−, kauranes
Wyethia	6	−, prenylated flavonones, $\equiv_5=$
Zexmenia	4	6.12 GE (1), 6.12 Eu (1), 8.12 GE (1), p-oxyacetoph. 8.12 E (10 α) (1)
Zinnia	13	6.12 Eu (10 α) (1), Gu (9), EL (10) (10), 6.12 Eu (2), 8.12 GE (3)
16. *Heliantheae* (Heli)		
Agiobampoa	1	−, bornylester
Garcilassa	2	−
Helianthopsis	3	FHe (3)
Helianthus	18	FHe (15), GE (19), kauranes $C_{17}\equiv$
Hymenostephium	1	−, $C_{17}\equiv$
Lagascea	2	Xa (1), oxyacetoph. $C_{17}\equiv$
Sclerocarpus	2	−
Simsia	3	8.12 GE (1), $C_{17}\equiv$, p-oxyacetoph.
Syncretocarpus	1	Ca (1), FHe (1), $C_{17}\equiv$
Tithonia	7	FHe (3), GE (6), Eu (2)
Viguiera	35	FHe (22), GE (15), kauranes
17. *Neurolaeninae* (Neuro)		
Brasilia	1	FHe (1), GE (1)
Calea	34	FHe (19), Gu (6), GE (19), Eu (5)
Greenmaniella	1	FHe (1), GE (1), Ge (1), Eu (1)
Neurolaena	4	FHe (1), GE (2)
18. *Galinsoginae* (Galins)		
Alloiospermum	1	−, $C_{17}\equiv$
Bebbia	1	−, $C_{17}\equiv$
Galinsoga	4	−, $C_{17}\equiv$
Jaegeria	1	−, $C_{17}\equiv$
Schistocarpha	5	−, $C_{17}\equiv$
Selloa	1	−, $C_{17}\equiv$
Tetragonotheca	3	ME (2), $C_{17}\equiv$
Tridax	5	−, $C_{17}\equiv$
19. *Dimeresiinae*	−	
20. *Coreopsidinae* (Coreops)		
Bidens	29	−, = $\equiv_4=$ a. o., phenylpropanes
Chrysanthellum	2	−
Coreopsis	42	−, = $\equiv_4=$ a. o., phenylpropanes
Cosmos	5	−, = $\equiv_4=$ a. o., phenylpropanes
Dahlia	11	−, = $\equiv_4=$ a. o.
Glossocardia	1	−, = $\equiv_4=$ a. o.
Glossogyne	1	−
Heterosperma	2	−, = $\equiv_4=$ a. o.
Isostigma	1	−, = $\equiv_4=$ a. o.
Thelesperma	4	−, = $\equiv_4=$
Venegasia	1	GE (1)
21. *Fitchiinae*		
Fitchia	1	Gu (1), phenylpropane, oxyacetophenones

Table 1 continued

Heliantheae	n	Main constituents
22. *Coulterellinae* (Coult)		
Coulterella	1	$-$, $\equiv_5=$, Th. $\equiv$, phenylpropanes
23. *Pectidinae* (Pect)		
Adenophyllum	1	$-$, Th. $\equiv$
Chrysactinia	1	$-$, Th. $\equiv$
Dyssodia	6	$-$, Th. $\equiv$
Hymenatherum	1	$-$
Pectis	7	GE (1), Gu (1), Eu (1), thymol der.
Porophyllum	5	$-$, $\equiv$, thymol der.
Schizotrichia	1	$-$, Th. $\equiv$
Tagetes	21	$-$, Th. $\equiv$
Thymophylla	1	$-$, Th. $\equiv$
24. *Flaveriinae* (Flav)		
Flaveria	10	$-$, Th. $\equiv$
25. *Varillinae*	$-$	
26. *Clappiinae*	$-$	
27. *Jaumeinae*		
Jaumea	1	$-$, $\equiv_5=$, Th. $\equiv$
28. *Madiinae* (Mad)		
Achyrachaena	1	$-$, $\equiv_5=$, $\equiv$
Calycadenia	1	$-$, $\equiv_5=$ $\equiv$
Hemizonia	10	$-$, $\equiv_5=$, oxyacetophenones, labdanes
Layia	3	$-$, $\equiv_5=$, Th. $\equiv$
Madia	3	$-$, $\equiv_5=$, oxyacetophenones, labdanes
29. *Hymenopappinae* (Hym)		
Hymenopappus	5	GE (3), Gu (4), $\equiv_5=$
Loxothysanus	1	Gu (1), PGu (1)
Villanova	1	GE (1)
30. *Lycaprinae*	$-$	
31. *Peritylinae*		
Perityle	1	$-$
32. *Baeriinae* (Baer)		
Amblyolepis	1	$-$
Baeria	4	Gu (2), Th. $\equiv$
Eriophyllum	3	GE (2), SEu (1), Eu (1)
Monolopia	1	$-$, $\equiv_5=$
Lasthenia	17	$-$, $\equiv_5=$, Th. $\equiv$
33. *Chaenactidinae* (Chaen)		
Arnica	16	PGu (9), Za (2), $\equiv_5=$, thymol der.
Bahia	2	Gu (2)
Chaenactis	3	GE (2), Th. $\equiv$
Hymenothrix	1	Gu (1), rosanes
Othake	1	$-$
Palafoxia	5	$-$, Th. $\equiv$, rosanes
Peucephyllum	1	GE (1)
Picradeniopsis	3	FHe (2), Gu (1), SGE (1), Th. $\equiv$
Schkuhria	10	GE (7), EL (3), Th. $\equiv$

Table 1 continued

Heliantheae	n	Main constituents
34. *Gaillardiinae* (Gaill)		
Baileya	3	PGu (3), GE (2), Gu (2)
Balduina	2	PGu (2)
Dugaldia	2	PGu (2), SPGu (2)
Gaillardia	17	PGu (16), Eu (4), Gu (3), Stu. ≡
Haploesthes	1	−, Th. ≡
Helenium	35	PGu (33), Gu (9), SPGu (1), Eu (1), Th. ≡, ≡ Stu. ≡
Hymenoxys	18	PGu (6), SPGu (6), Gu (1)
Plummera	2	−
Psilostrophe	3	SPGu (2), PGu (1)
35. *Marshalliinae* (Marsh)		
Marshallia	2	−, thymol der., pren. flavanones

from *Xanthium*. As these compounds are also present in the *Inuleae* (*Helichrysum, Achyrocline* and *Pterocaulon*) a relationship to the *Inuleae* may be supported.

The overall picture of *Heliantheae* chemistry indicates that the separation into the subtribes proposed is supported only in part by chemistry. While several tribes have a characteristic chemistry, others have not. The limits of the tribe *Ecliptinae* probably are too wide. Whereas sesquiterpene lactones have been reported from 16 subtribes, they seem to be absent in 11. The other characteristic group of compounds, the polyynes, derived from the pentaynene, are more or less present in all subtribes, though eight small ones have not been investigated so far.

References

ROBINSON, H., 1981: A revision of the tribal and subtribal limits of the *Heliantheae* (*Asteraceae*). − Smithsonian Contr. Bot. **51**: 1−102.

SEAMAN, F. C., 1982: Sesquiterpene lactones as taxonomic characters in the *Asteraceae*. − Bot. Rev. **48**: 201−214.

Address of the author: F. BOHLMANN, Department of Organic Chemistry, Technical University, Strasse des 17. Juni 135, D-1000 Berlin 12.

Pl. Syst. Evol. [Suppl. 4], 77–97 (1990)

Isolating mechanisms and implications for modes of speciation in *Heliantheae* (*Compositae*)

Scott D. Sundberg and Tod F. Stuessy

Received December 4, 1987

Key words: Angiosperms, *Compositae, Heliantheae, Melampodium.* – Isolating mechanisms, speciation.

Abstract: The *Heliantheae* contain more than 200 genera and 2500 species. To aid understanding of the origin of species within the tribe, a survey of the extant isolating mechanisms between and among closely related species has been undertaken using available monographs. The predominant isolating factor is reproductive isolation, with many closely related taxa differing in chromosome number or being incompatible. Geographical isolation was found in 67% of the groups. Differences in flowering time and habitat have been isolating factors in some instances. The existence of autogamy and pollinator preference as isolating mechanisms is uncertain due to scanty data. Significant differences in pollination systems seem unlikely due to widespread promiscuous pollination in the tribe. The general role of isolating mechanisms in speciation in the *Heliantheae* suggests geographic and allopolyploid modes as most common with recombinational and quantum speciation of lesser significance.

The *Heliantheae* constitute one of the largest tribes of the *Compositae*, with more than 200 genera and 2500 species. Recently the subtribal structure of the *Heliantheae* has been considered by STUESSY (1977 a) and ROBINSON (1981), and these studies have provided a much clearer view of the higher-level classification of the tribe. Further, many revisions and monographs have been produced in recent decades, providing in-depth analysis of relationships and including new information on distributions, chromosome numbers, phylogeny, and biosystematic data. Some of these studies have focused on evolutionary considerations, especially mechanisms of speciation, such as those in *Coreopsis* (SMITH 1974, CRAWFORD & SMITH 1982) and *Dubautia* (CARR & KYHOS 1981, CARR & al. 1986, CARR 1985). These detailed studies on particular genera, and the existing recent subtribal structure of the tribe, now provide the opportunity to examine aspects of the origin and evolution of taxa within the tribe.

Of various factors which might be considered in the evolution of *Heliantheae*, isolating mechanisms are most important for maintenance of diverse populational systems and as stimulus for speciation. The purposes of this paper, therefore, are to: (1) assess the predominance of different isolating mechanisms within the *Heliantheae*; (2) correlate the prevalence of isolating mechanisms with each other and with longevity and type of capitulum (discoid versus radiate heads); (3) determine if patterns detected in the general survey are reflected in particular genera for which detailed information is available; and (4) offer suggestions for modes of speciation within the tribe.

Materials and methods

A survey of available literature was conducted to identify groups (usually pairs) of closely related species in genera of the *Heliantheae* and to gather data on isolating mechanisms that maintain the genetic integrity of species within the groups. Literature surveyed included recent monographs, revisions, and other relevant biosystematic publications. When needed, supplementary information (e.g. distribution, phenology, chromosome counts) was obtained from modern floras or other sources. Species groups were included in the study only when the author expressly stated that the species were "closely related," had "close affinities," showed them as adjacent terminal taxa in phylograms or cladograms, or otherwise indicated that the species had close evolutionary affinities. Species that were compared using phrases such as "similar to," "morphologically close to," or others that expressed phenetic, but not necessarily evolutionary, similarity were exluded from the study. Species concepts varied among the treatments. In cases where the author employed particularly narrow species limits or expressed doubt as to the validity of a taxon's specific status, the species were omitted. Likewise, some species groups were excluded on the basis of scanty data. Distributional data that were inadequate for distinguishing between allopatric and sympatric patterns were omitted, although some of these may reflect parapatric distributions.

The data base for the literature surveys includes taxa in all of the subtribes of the *Heliantheae* (following STUESSY 1977 a): *Melampodiinae* (7 genera), *Zinniinae* (5), *Ecliptinae* (4), *Verbesininae* (8), *Helianthinae* (9), *Gaillardiinae* (4), *Coreopsidinae* (6), *Fitchiinae* (1), *Bahiinae* (4), *Madiinae* (6), *Galinsoginae* (11), *Neurolaeninae* (4), *Engelmanniinae* (2), *Ambrosiinae* (3), *Milleriinae* (1). Data on these taxa were obtained from the following references: ABRAMS & FERRIS 1960, ADSERSEN 1980, BAAGØE 1974, BECKER 1979, BLAKE 1921, CANNE 1977, 1983, CARLQUIST 1957, CARR 1975, 1985; CARTER 1964, CHANDLER & al. 1986, CHANNELL 1957, CLARK & KYHOS 1980, CLAUSEN 1951, COLEMAN 1966 a, b, 1968, 1971, 1974, 1977, CRAWFORD 1970, 1976, 1982, CRONQUIST 1980, DILLON 1984, DYAR 1959, ELIASSON 1974, ELLISON 1964, FAY 1978, FISHER 1957, 1961, FUNK 1982, GARDNER 1979, GRASHOFF & al. 1972, HART 1979, HARTMAN & STUESSY 1983, HEISER 1956, 1961, 1965, HEISER & al. 1969, HUMBLES 1972, JACKSON 1960, 1963, JANSEN 1985, KATZ & TORRES 1965, KEIL & PINKAVA 1976, KEIL & STUESSY 1975, 1977, LA DUKE 1982, LONGPRE 1970, LÖVE & DANSEREAU 1959, McGREGOR 1968, McVAUGH & SMITH 1967, MEARS 1973, 1975, MESFIN T. 1984, pers. comm., OLSEN 1979 a, b, PANERO 1986, PARKER 1962, PARKER & JONES 1975, PARKS 1973, PAYNE 1964, 1966, PETERSON & PAYNE 1973, PINKAVA 1967, PINKAVA & KEIL 1977, POWELL 1965, 1968, POWELL & POWELL 1968, POWELL & SIKES 1970, POWELL & al. 1974, 1975, RABAKONANDRIANINA 1980, ROBINSON 1979, ROCK 1975, ROLLINS 1950, SANDERS 1977, SCHILLING E. pers. comm., SCHILLING & HEISER 1981, SCHILLING & SCHILLING 1986, SEAMAN & MABRY 1979, SEMPLE 1978, SEMPLE & SEMPLE 1978, SETTLE & FISHER 1972, SMITH 1974, 1976, 1982, 1983, SORENSEN 1969, 1980; SPEESE & BALDWIN 1952, SPOONER D. pers. comm., ST. JOHN 1971, STROTHER 1976, 1979, 1983, STUESSY 1970, 1972, 1973, 1977 b, 1978, 1979, STUESSY & BRUNKEN 1979, SUNDBERG & al. 1986, TANOWITZ 1978, 1982, TORRES 1963 a, b, 1964 a, b, 1968, 1969, TURNER 1966, 1972 a, b, 1978 a, b, 1980, 1982 a, b, 1984, 1985, TURNER & DAVIES 1980, TURNER & DAWSON 1980, TURNER & MORRIS 1976, TURNER & POWELL 1977, URBATSCH & al. 1986, VAN FAASEN 1971, VENKATESH 1958, WEBER 1952, WELLS 1965, 1969, WHALEN 1977, WUSSOW & URBATSCH 1978, 1979, WUSSOW & al. 1985. A table containing the data set is available from the authors.

Calculations were based on data from species pairs with all relevant data. For example, the calculation of the co-occurrence of spatial isolation and isolation due to habitat difference was made from only those pairs with full data for both isolating mechanisms. Data on triplets of species were gathered in a similar fashion to those involving pairs of closely related taxa. All information regarding the predominance of isolating mechanisms and their correlations with each other show the same trends for triplets of species as they do for pairs, so the data were pooled. This yielded 52 triplets and 168 pairs of species analyzed in the survey, resulting in 324 pair-wise comparisons.

Outline of isolating mechanisms in angiosperms

Different classifications of isolating mechanisms exist (e.g., LEVIN 1978, GRANT 1981, 1985; WHITE 1978, LITTLEJOHN 1981). For purposes of this study, we have selected the outlines in GRANT (1981, 1985) as being most useful (Table 1 in part). To obtain a general idea of isolating factors within the *Heliantheae*, we have focused on those aspects for which data are routinely contained in monographs and revisions, namely spatial, environmental, temporal, and reproductive. Several factors such as diurnal, ethological, mechanical, and gametic isolation have not been reported in the tribe. Likewise, data are not adequate to distinguish pre- and post-fertilization barriers nor the specific causes for hybrid sterility (genic, chromosomal, or cytoplasmic). Table 1 shows those factors of significance in *Heliantheae* and also in two genera, *Melampodium* and *Calycadenia*, to be discussed in more detail later in the paper. Particularly detailed studies have been done on isolating mechanisms in these taxa, which serve to test the results from the general literature survey.

Survey of isolating mechanisms in *Heliantheae*

The survey of 324 pairs of species in the *Heliantheae* and their isolating mechanisms of necessity contains several omissions. The data pertaining to all the isolating mechanisms in each of the closely related species pairs are somewhat uneven. For

Table 1. Outline of isolating mechanisms (after GRANT 1981, 1985) in *Heliantheae* and their occurrence in *Melampodium* and *Calycadenia*. *c* Common, *r* rare, *n* not known to occur, *na* data not available

Isolating mechanisms in *Heliantheae*	*Melampodium*	*Calycadenia*
Spatial		
geographical isolation	c	c
Environmental		
ecological isolation	r	c
Reproductive		
premating barriers		
temporal isolation: seasonal	n	n
isolation due to autogamy	n	r
postmating barriers		
incompatibility barriers	c	c
hybrid inviability	n	na
hybrid sterility	r	c
hybrid breakdown	n	na

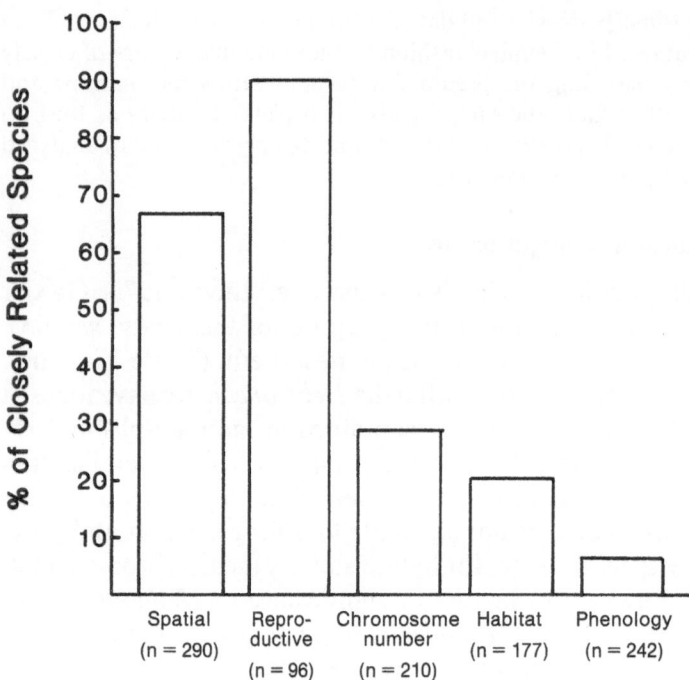

Fig. 1

example, although all included studies show distributions and discuss interspecific relationships, not many provide data from hybridization studies which bear upon postmating barriers. Ten percent had limited distributional data, 45% were missing adequate habitat data, 25% did not include information on flowering time (phenology), and 70% provided no indication of reproductive isolation (35% lacking chromosome numbers; 83% lacking hybridization data).

The predominant isolating mechanisms in *Heliantheae* are shown in Fig. 1. Reproductive and spatial isolation are the most common modes within closely related species pairs of the tribe. Differences in phenology are insignificant as isolating factors in the *Heliantheae*. Likewise, very little habitat differentiation can be seen that could account for isolation in closely related species pairs. Reproductive isolation comprises many different types of barriers, and a few comments seem appropriate. Chromosome numbers which are routinely reported in revisions and monographs show differences between closely related species in 28% of the cases in the survey. Of the types of reproductive isolation, the data are not uniform and therefore a careful percentage breakdown is impossible, but it is worth noting that many different types have been encountered (number of reports shown in parentheses): euploid chromosome number differences (46); aneuploid chromosome number differences (13); incompatibility barriers (12); hybrid inviability (1); hybrid sterility (26); hybrid breakdown (5); hybrids obtained but no further data (20), and a few cases in which no reproductive isolation was indicated (11).

There are few accounts of autogamy in the *Heliantheae*. In addition, there was seldom any indication whether the self-compatible species were obligate or facultative selfers. The self-compatible species included in the study are usually isolated from their close relatives by additional isolating mechanisms. Thus the insular, presumed self-compatible species, *Acmella darwinii* (PORTER) R. K. JANSEN and species of *Lipochaeta*, are allopatric to their respective relatives (JANSEN 1985, RABAKONANDRIANINA 1980). *Galinsoga quadriradiata* RUIZ & PAVON, a self-compatible species, is reproductively isolated from its closest relatives, as evidenced by differences in ploidy level (CANNE 1977). *Parthenium argentatum* GRAY and *P. incanum* KUNTH, which are facultatively autogamous, are normally isolated by differences in phenology and although occasional hybrids are found in nature, there are no intermediate populations known (ROLLINS 1950). *Polymnia canadensis* L. and its presumed derivative, *P. laevigata* BEADLE, are moderately self-compatible, with self-pollinated seed sets of about 9 and 55%, respectively (WELLS 1969). The two species have overlapping ranges, but are not known to occur together or hybridize in nature. Artificial hybrids between the species have high pollen stainability (89%, WELLS 1969). The self-compatible species, *Calycadenia hooveri* CARR is reportedly a derivative of the outcrossing species *C. villosa* DC. (CARR 1975). The two species are allopatric and artificial hybrids have meiotic irregularities, an indicator of at least partial reproductive isolation. Obligate outcrossing was reported for at least some species of *Balduina, Bidens, Coreopsis, Guardiola, Helianthus, Simsia, Viguiera, Xanthium, Zinnia*, and in much of the *Madiinae* (PARKER & JONES 1975, HART 1979, VAN FAASEN 1971, SMITH 1976, HEISER & al. 1969, SPOONER D. pers. comm., SCHILLING E. pers. comm., HICKS 1975, OLORODE 1970, TORRES 1963 b, CARR & al. 1986, respectively). Since most species of *Heliantheae* have showy floral displays and typically lack characteristics often associated with autogamy (e.g., inconspicuous floral presentation, small heads, etc., ORNDUFF 1969) it is likely that the breeding system that predominates in the *Heliantheae* is that of obligate outcrossing or at most, facultative selfing. Autogamy apparently plays a minor role in reproductive isolation in the *Heliantheae*.

In the *Heliantheae* co-occurrence of some of the isolating mechanisms separating closely related taxa is seen (Table 2). The co-occurrence of two or more isolating mechanisms is low in general, which suggests that a single isolating mechanism is normally adequate for maintaining species integrity. However, a 45% correlation was found between spatial and reproductive isolation (including chromosome number differences). The low correlation between habitat and spatial isolation indicates that differences in distribution are not necessarily correlated with changes in habitat. It may well be that microhabitat differences do occur and have not been documented in the literature, but certainly marked ecological differences are not common (but see discussion below). Habitat differences are not highly correlated with reproductive isolating factors either (13%). Phenological differences do not correlate

Table 2. Co-occurrence of isolating mechanisms between species in *Heliantheae*, shown as percent of sample for which data pertain to both isolating mechanisms

	Spatial (%)	Reproductive (%)	Habitat (%)
Reproductive	45		
Habitat	14	13	
Phenology	4	4	1

Table 3. Comparison of percentages of isolating mechanisms occurring in allopatric and sympatric closely related species in *Heliantheae*. Sample sizes shown in parentheses

	Allopatric (%)	Sympatric (%)
Reproductive	88 (41)	92 (39)
Chromosome number	22 (113)	37 (68)
Habitat	23 (95)	12 (58)
Phenology	6 (136)	7 (81)

with any of the other factors, which may relate to the permiscuous pollination system within the *Compositae* as a whole and the *Heliantheae* in particular.

Because distributional patterns have an important role to play in isolation of related species, it is of interest to compare the relationship of allopatric and sympatric patterns with the other isolating factors (Table 3). Chromosome number differences are found more often in sympatric species (37%) than they are in allopatric ones (22%). Reproductive isolation (including chromosome number differences) also shows a higher difference between sympatric closely related species

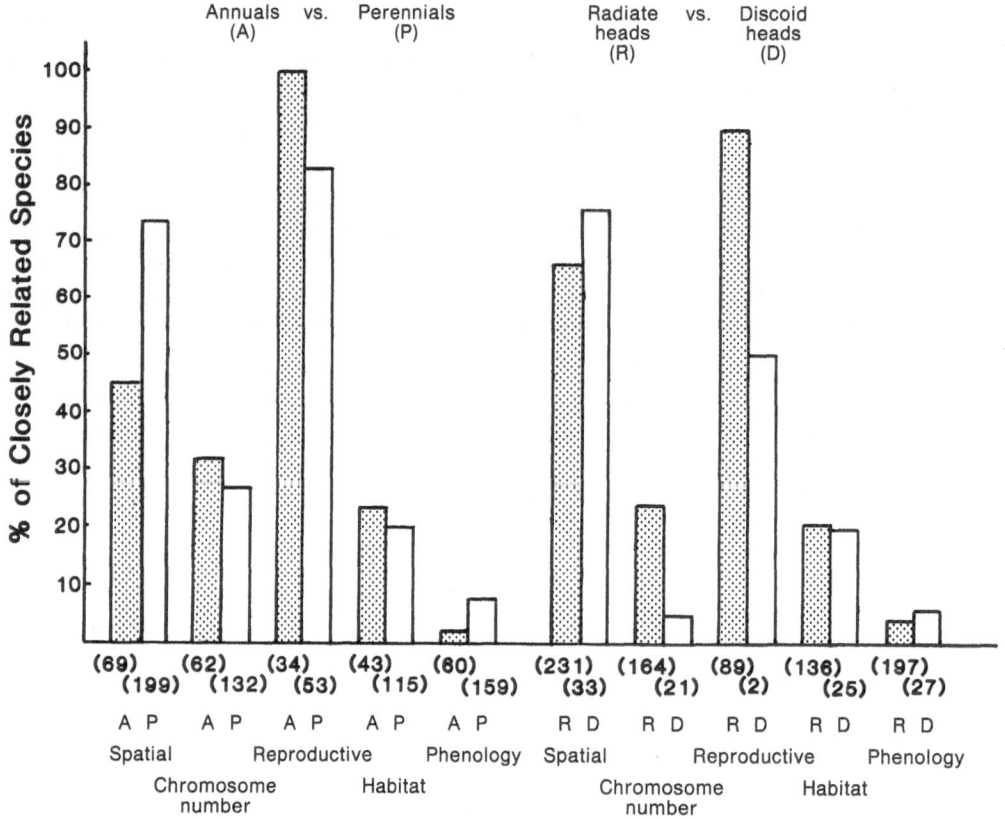

Fig. 2

than among allopatric ones. These results would be expected as spatial isolation obviates other mechanisms. Habitat differences occur twice as frequently in taxa that are allopatric (23 versus 12%), rather than sympatric. Ecological differentiation, therefore, can be an important correlate of allopatric speciation, although not commonly so. Phenological differences separate only a few species (6 – 7%) in either distributional configuration.

Also of interest is the correlation (Fig. 2) of types of isolating mechanisms with longevity (annual versus perennial) and type of flowering head (radiate versus discoid). Perennials are more spatially isolated than annuals. Explanation of this pattern lies in considering the chromosomal differences in relationship to the spatial patterns. Shrubby or tree-like perennials have an average of 1.8 different chromosome numbers per genus whereas herbaceous annuals and perennials have an average of 4.0 different chromosome numbers per genus (data calculated from STUESSY 1977, as the mean number of different chromosomal levels in genera with five or more species). Once speciation within perennial groups occurs and the taxa establish themselves allopatrically, they tend to remain that way due to longer generation times, lower levels of chromosomal modification, and inability to pass tests of sympatry with related taxa. This trend is further seen in the comparison between annuals and perennials and chromosomal differences that reflect isolating mechanisms. 32% of the annual species pairs surveyed show chromosomal differences but only 27% of the perennials show this trend. Annuals tend to be more reproductively isolated overall (including chromosomal number differences) than perennials. This supports the hypothesis that there is a greater selective advantage for the establishment of reproductive isolation in annual species (GRANT 1981). Isolation due to habitat differentiation does not appear to differ between annuals and perennials. Perennials tend to show more isolation due to flowering time than do annuals, although both are at low levels. Because perennials tend to have fewer reproductive differences in general, when they remain sympatric, differences in flowering time provide needed isolation.

In comparing radiate and discoid heads, the sample size for species pairs which are discoid is generally low. A difference in spatial isolation in radiate versus discoid groups occurs, but the difference (10%) is not easily explainable a priori. A conspicuous difference exists in reproductive isolation (also in chromosome number) examined separately, and these results may relate in some way to ethological factors involving pollinators and breeding systems. Radiate heads are known to attract more pollinators (STUESSY & al. 1986), which suggests that reproductive isolation between radiate taxa might be more imporant in the face of stronger pollinator visits. Also, some of the discoid taxa may be autogamous. There are no visible differences in habitat or phenological isolation in radiate versus discoid representatives.

Detailed studies of isolating mechanisms in selected genera

The previous survey shows general trends and patterns of isolating mechanisms within the *Heliantheae*. This survey is by nature broad in scope; it would therefore be useful to examine selected genera which reflect a broad range of isolating mechanisms to determine if the same correlations and trends found in the survey prevail within well-studied groups. Many genera might be selected for detailed

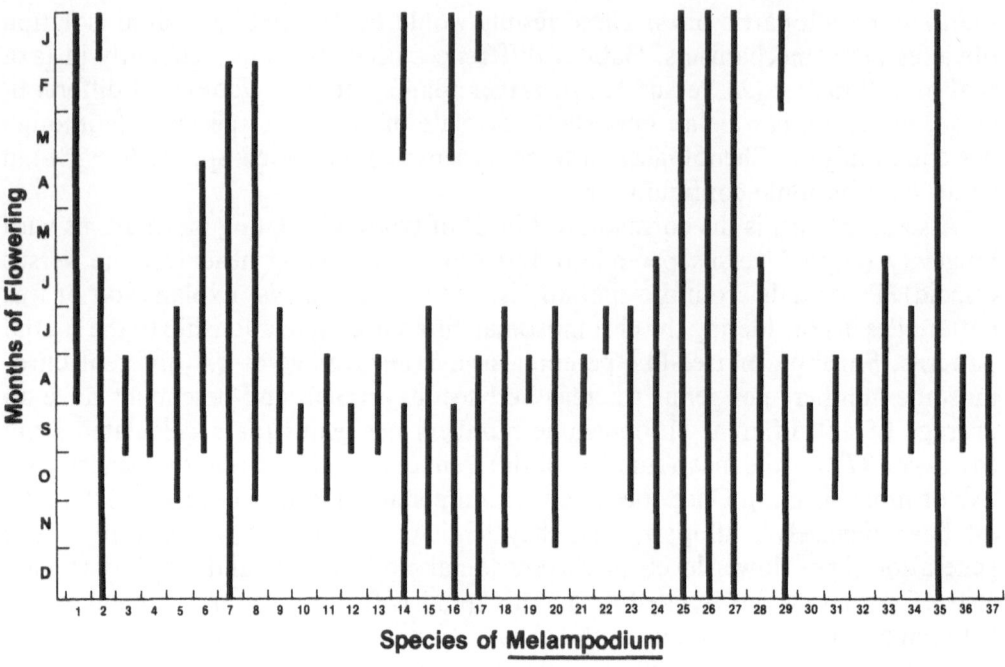

Fig. 3

examination but two seem appropriate for our needs: *Melampodium* (subtr. *Melampodiinae*) and *Calycadenia* (*Madiinae*). *Melampodium* contains a remarkable chromosomal series, and geological, habitat and phenological data are available. *Calycadenia* contains much information on reproductive isolation, breeding systems, and phenology.

Melampodium contains 37 species distributed throughout Mexico and Central America (Stuessy 1972). The genus has been the subject of several previous investigations including revisionary studies (Stuessy 1971 a, 1972), cytology (Stuessy 1971 b), phytochemistry (Fischer & al. 1972), cladistics (Stuessy 1979), phenetics (Stuessy & Crisci 1984), and a limited number of interspecific hybridizations (Stuessy & Brunken 1979). For this discussion data on cytology and distribution of species have been taken from Stuessy (1971 b, 1972). In addition, data on sympatric occurrence of species have been obtained from personal field notebooks of trips to Mexico and Central America during 1965 and 1966 which totaled three months and over 500 collections.

Of the known isolating mechanisms in *Melampodium* (Table 1), geographical isolation and incompatibility barriers are most common. Isolation due to autogamy and seasonal isolation are unknown. Ecological isolation and that due to hybrid sterility are rare, although present within the genus (the former exists between sections rather than between closely related species). Most species of *Melampodium* flower during August and September, although some species have a much broader range of flowering throughout the year (Fig. 3). These species are more weedy, are found in different habitat zones, and survive throughout different seasons. Nonetheless, phenological isolation is not common within the genus. Habitat differentiation has occurred to a marked degree within *Melampodium* (Fig. 4). Seven prin-

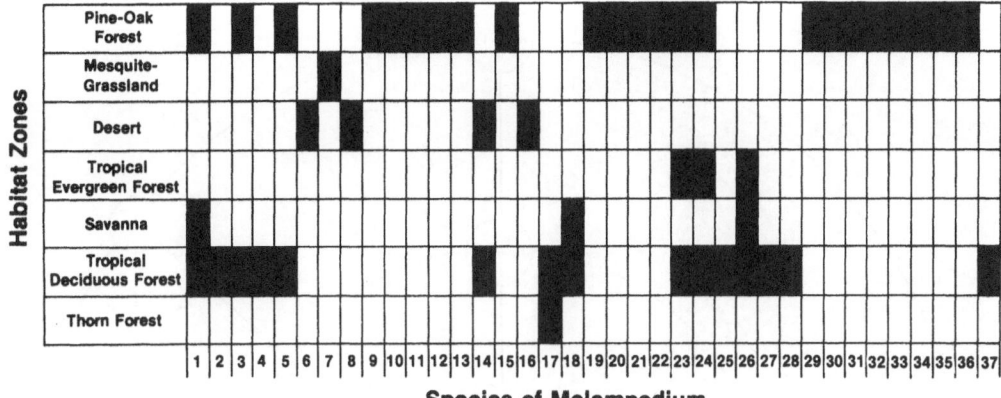

Fig. 4

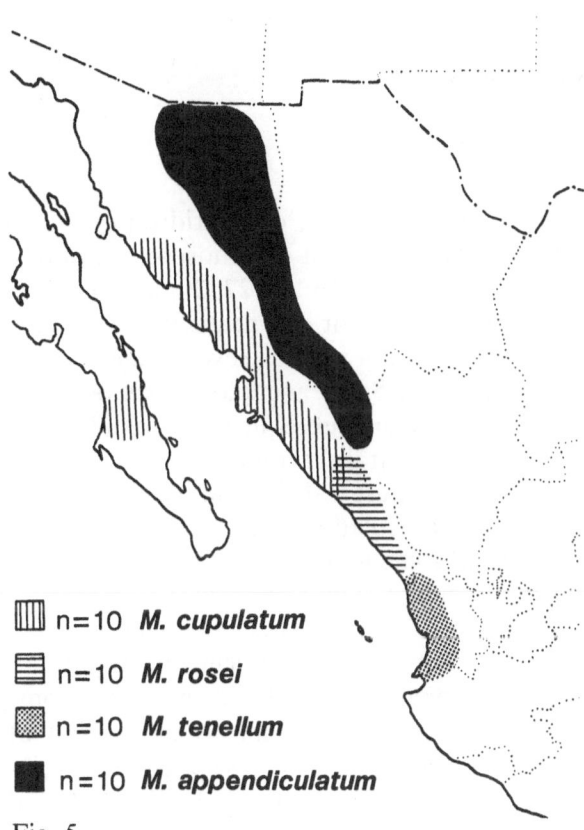

⊞ n=10 *M. cupulatum*

⊟ n=10 *M. rosei*

▦ n=10 *M. tenellum*

■ n=10 *M. appendiculatum*

Fig. 5

cipal habitat categories have been recognized in which the species of the genus are known to occur. These include the major habitat zones of Mexico and Central America (after LEOPOLD 1950). Even more significantly, spatial isolation is a conspicuous pattern within the genus. Closely related species almost always have allopatric distributions, as shown in the example of ser. *Cupulata* from northwestern Mexico (Fig. 5). Where closely related species do overlap parapatrically, morpho-

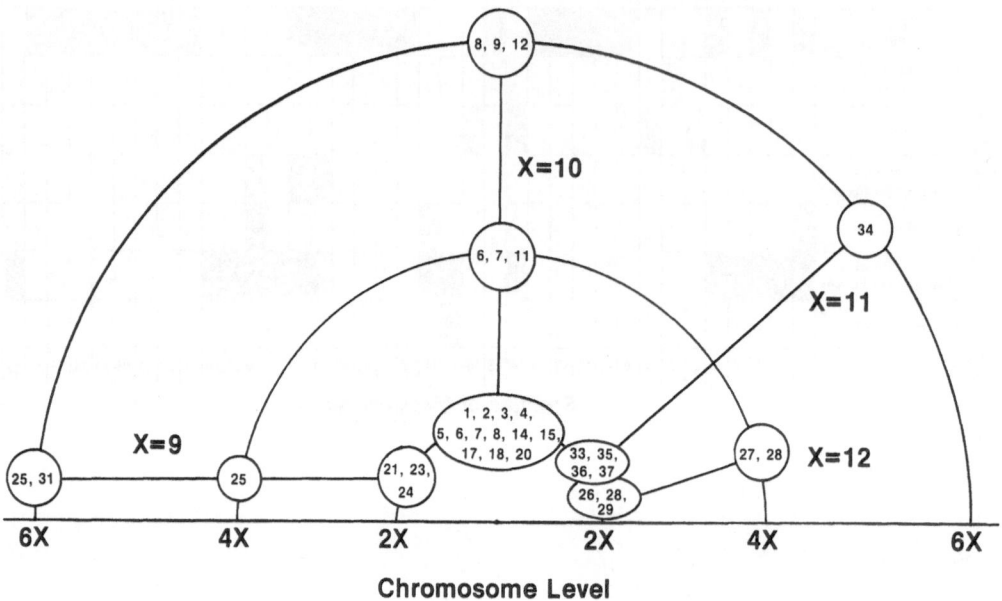

Fig. 6

logical intermediates occur which suggest the occurrence of hybridization. Even more significantly in *Melampodium* is the broad range of chromosomal variation with haploid numbers of n = 9, 10, 11, 12, 18, 20, 23, 25 ± 1, 27, 30, and 33. This is one of the longest series of chromosome numbers for any genus in the tribe. It is likely that the chromosomal system has evolved in parallel several times within the genus (Fig. 6) based on x = 9, 10, 11, and 12. Some aneuploidy has occurred from the original chromosomal lines as well as from subsequent polyploid levels. The pattern, then, is of a complex development of parallel evolution in chromosome numbers with polyploidy being conspicuous. Because habitat and elevational differences do correlate with geographic patterns of isolation, there is an opportunity to examine directly the relationship of chromosome numbers and spatial isolation.

Personal field observations reveal the common sympatric occurrence of 2 − 4 species of *Melampodium* in any single locality (Table 4). These are often growing together in a radius of several meters in what appears to be the same microhabitat. Most important is that the species found together always have different chromosome numbers. Sympatry within *Melampodium* has occurred primarily as a result of broad changes in chromosome number within the genus. In closely related species with different chromosome numbers, they tend to have overlapping distributional patterns (e.g., ser. *Sericea*, Fig. 7). *Melampodium*, therefore, provides an example of a genus that is isolated primarily by a combination of spatial and chromosomal factors.

Calycadenia is a genus of about 12 annual species that range from southern Oregon to Baja California, Mexico and are most diverse in central California. Several evolutionary studies of the genus have focused on hybridization experiments and cytological observations (Carr 1975, 1976, 1977, 1980; Carr & Carr 1982, 1983); and an investigation has centered on the flavonoid constituents of one species (Emerson & al. 1986). Data for the following discussion were obtained from these sources and from KECK (1960). Fifty-five paired combinations among 11 species of *Calycadenia* were analyzed (omitting *C. fremontii* GRAY due to lack of information).

Speciation within *Calycadenia* has occurred at the diploid level and no polyploidy has been reported. It has involved aneuploid changes and chromosomal repatterning, with or without marked morphological differentiation (CARR 1977). A consequence of these extensive structural rearrangements of the chromosomes has been the development of strong reproductive isolation between the species, as seen in hybridization experiments (CARR 1975, 1976, 1977) and as inferred from observations of differences in chromosome numbers. Of the 55 paired combinations of species of *Calycadenia* data bearing on reproductive isolation are known for 51. With the exception of a single pair, these are all reproductively isolated by differences in chromosome number, hybrid sterility, or incompatibility barriers. The exception is a pair of species (*C. multiglandulosa* DC. and *C. hispida* E. GREENE) that: are cytologically uniform with respect to chromosomal arrangement, are known to hybridize in nature, and yield artificial hybrids with high pollen stainability (68%) and consistent association of chromosomes in bivalents during meiosis (CARR 1977). Nothing has been reported either on the extent of gene flow between the species in areas of sympatry or on the possibility of isolation due to hybrid breakdown, however, and thus they may still be reproductively isolated. Isolation due to autogamy may contribute to maintaining the genetic integrity of *Calycadenia hooveri*, although it is also reproductively isolated in other ways (CARR 1975).

Table 4. Observed sympatry of species of *Melampodium* without known hybridization. Data from author's field notebooks of 1965 and 1966. Numbers in the data matrix refer to the number of times the pairs of species were found together; numbers in parentheses refer to haploid chromosome numbers. M. am. *M. americanum*, M. dic. *M. dicoelocarpum*, M. div. *M. divaricatum*, M. gl. *M. glabrum*, M. gr. *M. gracile*, M. li. *M. linearilobum*, M. lo. *M. longipes*, M. lon. *M. longipilum*, M. mi. *M. microcephalum*, M. pa. *M. paniculatum*, M. pe. *M. perfoliatum*, M. ro. *M. rosei*, M. se. *M. sericeum*

	M. am. (10)	M. dic. (12)	M. dic. (23)	M. div. (12)	M. gl. (11)	M. gr. (9)	M. li (10)	M. lo. (10)	M. lon. (10)	M. mi. (9)	M. pa. (18)	M. pe. (11)	M. ro. (10)	M. se. (30)
M. dicoelocarpum (12)	1													
M. dicoelocarpum (23)	–	–												
M. divaricatum (12)	2	–	1											
M. glabrum (11)	–	–	–	–										
M. gracile (9)	1	1	–	5	–									
M. linearilobum (10)	–	–	–	1	–	–								
M. longipes (10)	–	–	–	–	–	–	–							
M. longipilum (10)	–	–	–	1	–	–	–	–						
M. microcephalum (9)	1	–	–	1	–	–	–	–	–					
M. paniculatum (18)	–	–	–	3	–	–	–	–	–	–				
M. perfoliatum (11)	1	–	1	9	–	–	–	–	–	–	1			
M. rosei (10)	–	–	–	1	–	–	–	–	–	–	–	–		
M. sericeum (30)	2	–	1	9	2	1	1	1	–	1	–	5	–	
M. tenellum (10)	–	–	–	1	–	–	–	–	–	–	–	–	–	–

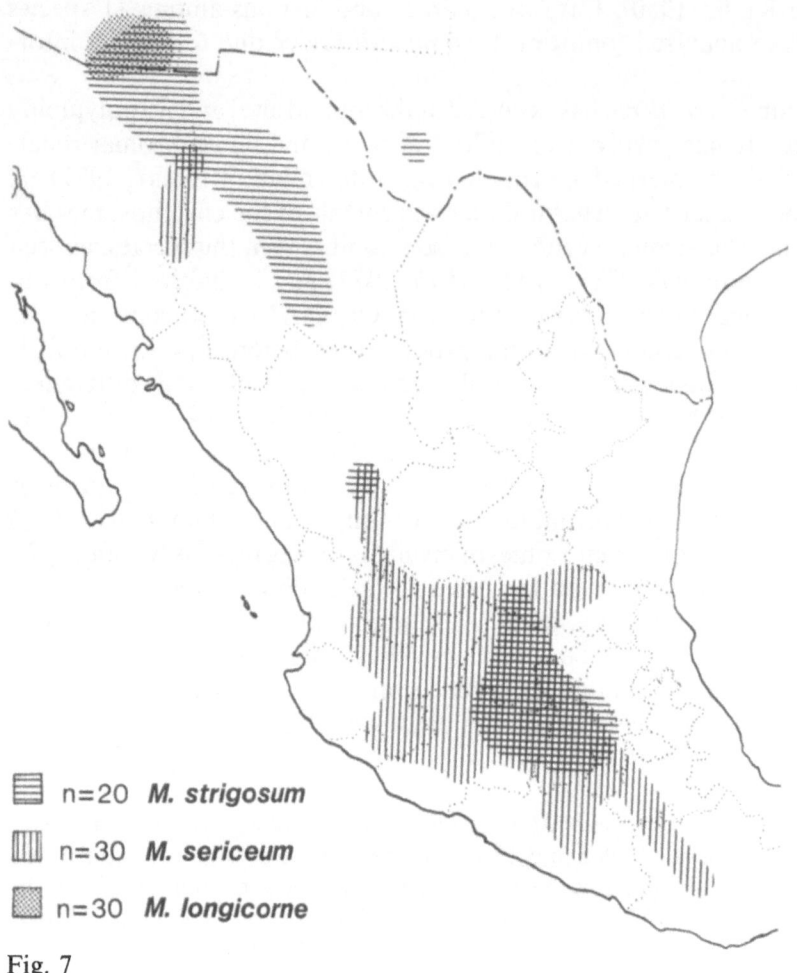

n=20	*M. strigosum*	
n=30	*M. sericeum*	
n=30	*M. longicorne*	

Fig. 7

Detailed distributional data, which are needed to determine the occurrence of allopatry and sympatry between species are not available for half of the species of *Calycadenia*. Nevertheless, general trends can be detected. Species pairs with allopatric distributions outnumber those that are sympatric by a large ratio (approximately 7:1 among pairs with adequate data). Closely related species of *Calycadenia* tend to be allopatric (or parapatric). Habitat differentiation may play a major role in maintaining isolation between species, as suggested by the predominance of allopatry and the wide range of floristic provinces occupied by species in the genus, but little specific information is available. In areas of sympatry, microhabitat differences are seen between populations of *Calycadenia hispida* and *C. multiglandulosa*, and between *C. pauciflora* GRAY and *C. ciliosa* E. GREENE, respectively. Phenological isolation is of little significance in the genus. Indeed, there is considerable overlap in flowering times among the 11 species studied.

In *Calycadenia* isolating mechanisms frequently act in concert to effect genetic isolation. This is most commonly seen as a combination of reproductive isolation and spatial isolation but may also involve habitat differentiation and isolation due to autogamy. It should be noted that hybrids between several pairs of species have been obtained by artificial hybridization (CARR 1977) and although at low rates, limited gene flow may potentially occur between sympatric species of flowering plants whose reproductive isolation is expressed at the postmating stage.

Implications for modes of speciation

Several well-established modes of speciation have been documented within flowering plants (after GRANT 1981) (Table 5). In the development of these different modes of speciation, the origin and maintenance of isolating mechanisms are fundamental. Some isolating mechanisms have arisen first and others have developed later as a byproduct of divergence and as a consequence of selective forces that subsequently arose. Studies on existing isolating mechanisms between species pairs will reveal not only first genetic isolating events but also those which developed subsequently. The first genetic isolating event is the most relevant to modes of speciation (WHITE 1978). However, in the recent and rapidly evolving genera of *Heliantheae* it seems likely that many observed isolating mechanisms do actually reflect the first genetic isolating events. Therefore, one can examine existing isolating mechanisms for suggestions on modes of speciation.

In the *Heliantheae* the principal isolating mechanisms existing among species pairs are reproductive and spatial. Isolation due to habitat and flowering time differences are of lesser importance. Taking into account the knowledge of predominance of these isolating mechanisms in the tribe, therefore, we can examine

Table 5. Correlation of principal isolating mechanisms with likely modes of speciation (adapted from GRANT 1981, 1985), and relative occurrence of each mode in *Heliantheae*. Small sample includes those species pairs for which full geographic, chromosomal, and reproductive hybridization data are available; large sample also contains species pairs with only geographic and chromosomal information. + Positive correlation; − negative correlation

Modes of speciation	Isolating mechanisms					Percent of *Heliantheae*	
	spatial	reproductive	chromosome number		autogamy	small sample	large sample
			aneuploid	euploid			
Geographical	+	−	−	−	−	5	52
Geographical or quantum	+	+	−	−	−	26	6
Quantum (involving aneuploidy)	+	+	+	−	−	8	3
Allopolyploid	+ or −	+	−	+	−	18	28
Recombinational (including aneuploidy)	−	+	+ or −	−	−	42	10
Selffertilization	+ or −	−	−	−	+	<1	<1

correlations with modes of speciation and their presumptive importance within the tribe (Table 5). Two different sample sizes were used, a smaller sample of 38 species pairs for which spatial, reproductive and chromosomal data are available, and a larger one of 164 species pairs also including examples for which only spatial and chromosomal data are known. Examination of both samples reveals evolution by autogamy, either in sympatric or allopatric populations, to be virtually unknown. Speciation involving allopolyploidy, a mode known to be common in angiosperms generally, is also common here, at levels of 18 and 28% in the two samples. Speciation involving aneuploidy in allopatric populations, labelled a type of quantum speciation here, is uncommon in the small and large samples at 8 and 3%, respectively. The other three modes of speciation are shown at very different levels in the two samples and therefore deserve more detailed comment. It is believed that the high levels of geographic speciation indicated in the large sample (52%), in contrast to the lower 5% in the smaller and more complete sample, reflects to some extent absence of reproductive data as reproductive isolation surely exists between many taxa separated geographically. Although many taxa speciate geographically, through time genetic divergence will usually occur as a result of different selection regimes in different environments. The smaller sample including only species pairs which have been extensively experimentally crossed stresses this point with the lower 5% value for geographic isolation being the only factor between closely related taxa. On the other hand, this value is surely somewhat artificially low and the 26% for geographic and quantum speciation artificially high due to overestimation of reproductive isolation because of the difficulty in crossing studies of showing no isolating factors of any kind. Further, investigators tend to cross extensively those taxa for which hybrids can be formed rather than attempt many crosses among taxa which would not hybridize under any circumstances. Nonetheless, these data are consistent with the concept of quantum speciation as suggested by GRANT (1985) to refer to a peripheral population which undergoes chromosomal changes accompanied by distribution to a new ecological setting. This divergence is accelerated by the founder effect accompanied by new selection pressures. Sympatric speciation involving the establishment of reproductive barriers appears common at 42 and 10%, respectively, but it is likely that many, if not most, of these examples reveal secondary rather than primary sympatry. That is, many of these examples may have originated via quantum or geographic speciation initially, developed reproductive isolation secondarily, and survived more recent sympatry. Therefore, taking all results into account, it seems that geographic and allopolyploid speciation are the most common modes in the *Heliantheae* with quantum and recombinational speciation of lesser importance.

Conclusion

The preceding surveys and discussions have summarized available information on isolating mechanisms and possible modes of speciation in the *Heliantheae*. We recognize that the only complete way to understand modes of speciation is to do investigations at the populational level. Here detailed experimental studies can sometimes reveal the crucial isolating mechanisms and subsequent morphological and genetic changes which have resulted in the origin of new taxa (e.g., GOTTLIEB 1973). Nonetheless, we believe strongly that an overview of these phenomena within the *Heliantheae* (with some general implications for the entire family) may serve as a stimulus for monographic workers to examine possible mechanisms for the origin of diversity in their groups more carefully. We hope this results in a more detailed and precise view of relationships within particular taxonomic groups and will serve as a stimulus for more intensive investigations by evolutionary biologists.

Appreciation is expressed to: The Ohio State University Graduate School for a Post-doctoral Fellowship in support of the senior author during the academic year, 1986 – 87; The Ohio State University College of Biological Sciences and the Department of Botany for funds to travel to the International Botanical Congress in Berlin for presentation of some of the results of this study; to Drs E. P. SCHILLING, D. SPOONER, and T. MESFIN for unpublished data on *Viguiera, Simsia,* and *Bidens* respectively; and to D. J. CRAWFORD for a critical reading of the manuscript.

References

ABRAMS, L., FERRIS, R. S., 1960: Illustrated flora of the Pacific states **4**. – San Jose: Stanford University Press.

ADSERSEN, H., 1980: Revision of the Galapagos endemic genus *Lecocarpus* (*Asteraceae*). – Bot. Tidsskr. **75**: 63 – 76.

BAAGØE, J., 1974: The genus *Guizotia* (*Compositae*). A taxonomic revision. – Bot. Tidsskr. **69**: 1 – 39.

BECKER, K. M., 1979: A monograph of the genus *Lasianthaea* (*Asteraceae*). – Mem. New York Bot. Gard. **31**: 1 – 64.

BLAKE, S. F., 1921: Revision of the genus *Acanthospermum*. – Contr. U.S. Natl. Herb. **20**: 383 – 392.

CANNE, J. M., 1977: A revision of the genus *Galinsoga* (*Compositae: Heliantheae*). – Rhodora **79**: 319 – 389.

– 1983: Cytological and morphological observations in *Galinsoga* and related genera (*Asteraceae*). – Rhodora **85**: 355 – 366.

CARLQUIST, S., 1957: The genus *Fitchia* (*Compositae*). – Univ. Calif. Publ. Bot. **29**: 1 – 144.

CARR, G. D., 1975: *Calycadenia hooveri* (*Asteraceae*), a new tarweed from California. – Brittonia **27**: 136 – 141.

– 1976: Chromosome evolution and aneuploid reduction in *Calycadenia pauciflora* (*Asteraceae*). – Evolution **29**: 681 – 699.

– 1977: A cytological conspectus of the genus *Calycadenia* (*Asteraceae*): an example of contrasting modes of evolution. – Am. J. Bot. **64**: 694 – 703.

– 1980: Experimental evidence for saltational chromosome evolution in *Calycadenia pauciflora* GRAY (*Asteraceae*). – Heredity **45**: 107 – 112.

– 1985: Monograph of the Hawaiian *Madiinae* (*Asteraceae*): *Argyroxiphium, Dubautia,* and *Wilkesia*. – Allertonia **4**: 1 – 123.

– POWELL, E. A., KYHOS, D. W., 1986: Self-incompatibility in the Hawaiian *Madiinae* (*Compositae*): an exception to Baker's Rule. – Evolution **40**: 430 – 434.

CARR, R. L., CARR, G. D., 1982: Micro- and nucleolar-organizing B-chromosomes in *Calycadenia* (*Asteraceae*). – Cytologia **47**: 79 – 87.

– – 1983: Chromosome races and structural heterozygosity in *Calycadenia ciliosa* GREENE (*Asteraceae*). – Am. J. Bot. **70**: 744 – 755.

– KYHOS, D. W., 1981: Adaptive radiation in the Hawaiian silversword alliance (*Compositae-Madiinae*). II. Cytogenetics of artificial and natural hybrids. – Evolution **40**: 959 – 976.

CARTER, A., 1964: The genus *Alvordia* (*Compositae*) of Baja California, Mexico. – Proc. Calif. Acad. Sci. **30**: 157 – 174.

CHANDLER, J. M., JAN, C.-C., BEARD, B. H., 1986: Chromosomal differentiation among the annual *Helianthus* species. − Syst. Bot. **11**: 354−371.

CHANNEL, R. B., 1957: A revisional study of the genus *Marshallia* (*Compositae*). − Contr. Gray Herb. **181**: 41−132.

CLARK, C., KYHOS, D. W., 1980: Specific status for *Encelia californica* var. *asperifolia* (*Compositae: Heliantheae*). − Madroño **27**: 48.

CLAUSEN, J., 1951: Stages in the evolution of plant species. − Ithaca: Cornell University Press.

COLEMAN, J. R., 1966 a: A taxonomic revision of section *Ximenesia* of the genus *Verbesina* L. (*Compositae*). − Am. Midl. Naturalist **76**: 475−481.

− 1966 b: A taxonomic revision of section *Sonoricola* of the genus *Verbesina* L. (*Compositae*). − Madroño **18**: 129−160.

− 1968: A cytotaxonomic study in *Verbesina* (*Compositae*). − Rhodora **70**: 95−102.

− 1971: The status of the genus *Actinomeris* NUTT. (= *Verbesina* L.) as revealed by experimental hybridization. − Bull. Torrey Bot. Club **98**: 327−331.

− 1974: *Verbesina* section *Ximenesia* (*Compositae*): biosystematics and adaptive radiation. − Am. J. Bot. **61**: 25−35.

− 1977: A summary of experimental hybridization in *Verbesina* (*Compositae*). − Rhodora **79**: 17−31.

CRAWFORD, D. J., 1970: Systematic studies on Mexican *Coreopsis* (sect. *Anathysana*), with special reference to the relationship between *C. cyclocarpa* and *C. pinnatisecta*. − Bull. Torrey Bot. Club **97**: 161−167.

− 1976: Taxonomy of *Coreopsis* sect. *Pseudo-Agarista* (*Compositae*) in Mexico with additional comments on sectional relationships in Mexican *Coreopsis*. − Brittonia **18**: 329−336.

− 1982: Chromosome numbers and taxonomic notes for Mexican *Coreopsis*, sections *Electra* and *Pseudoagarista*. − Brittonia **34**: 384−387.

− SMITH, E. B., 1982: Allozyme variation in *Coreopsis nuecensoides* and *C. nuecensis* (*Compositae*), a progenitor-derivative species pair. − Evolution **36**: 379−386.

CRONQUIST, A. C., 1980: Vascular flora of the southeastern United States. 1. *Asteraceae*. − Chapel Hill: University of North Carolina Press.

DILLON, M. O., 1984: A systematic study of *Flourensia* (*Asteraceae, Heliantheae*). − Fieldiana, Bot., n. s., **16**: 1−66.

DYAR, M. H., 1959: A biosystematic study of the genus *Berlandiera*. − Masters Thesis. Columbus: The Ohio State University.

ELIASSON, U., 1974: Studies in Galapagos plants. 14. The genus *Scalesia*. − Opera Bot. **36**: 1−117.

ELLISON, W. L., 1964: A systematic study of the genus *Bahia* (*Compositae*). − Rhodora **66**: 67−86, 177−215, 281−311.

EMERSON, J. K., CARR, R. L., McCORMICK, S., BOHM, B. A., 1986: 8-*o*-methylated flavones from *Calycadenia ciliosa* (*Compositae*): inter- and intrapopulational variation. − Biochem. Syst. Evol. **14**: 29−32.

FAY, J. J., 1978: Revision of *Perymenium* (*Asteraceae-Heliantheae*) in Mexico and Central America. − Allertonia **1**: 235−296.

FISCHER, N. H., WILEY, R., WANDER, J. D., 1972: Melampodin, a new germacranolide from *Melampodium leucanthum* TORR. & GRAY var. *leucanthum* (*Compositae*). − J. Chem. Soc., Chem. Comm. 1972: 137−139.

FISHER, T. R., 1957: Taxonomy of the genus *Heliopsis* (*Compositae*). − Ohio J. Sci. **57**: 171−191.

− 1961: A new species of *Heliopsis* from Mexico. − Ohio J. Sci. **61**: 178−179.

FUNK, V. A., 1982: The systematics of *Montanoa* (*Asteraceae, Heliantheae*). − Mem. New York Bot. Gard. **36**: 1−133.

GARDNER, R. C., 1979: Revision of *Lipochaeta* (*Compositae: Heliantheae*) of the Hawaiian Islands. — Rhodora **81**: 291–343.

GOTTLIEB, L. D., 1973: Genetic differentiation, sympatric speciation, and the origin of a diploid species of *Stephanomeria*. — Am. J. Bot. **60**: 545–553,

GRANT, V., 1981: Plant speciation. 2nd edn. — New York: Columbia University Press.

— 1985: The evolutionary process. A critical review of evolutionary theory. — New York: Columbia University Press.

GRASSHOFF, J. L., BIERNER, M. W., NORTHINGTON, D. K., 1972: Chromosome numbers in North and Central American *Compositae*. — Brittonia **24**: 379–394.

HART, C. R., 1979: The systematics of the *Bidens ferulaefolia* complex (*Compositae*). — Syst. Bot. **4**: 130–147.

HARTMAN, R. L., STUESSY, T. F., 1983: Revision of *Otopappus* (*Compositae, Heliantheae*). — Syst. Bot. **8**: 185–210.

HEISER, C. B., 1956: Biosystematics of *Helianthus debilis*. — Madroño **13**: 145–176.

— 1961: Morphological and cytological variation in *Helianthus petiolaris* with notes on related species. — Evolution **15**: 245–258.

— 1965: Species crosses in *Helianthus:* 3. Delimitation of "sections." — Ann. Missouri Bot. Gard. **52**: 364–370.

— SMITH, D. M., CLEVENGER, S. B., MARTIN, W. C., 1969: The North American sunflowers (*Helianthus*). — Mem. Torrey Bot. Club **22**: 1–218.

HICKS, A. J., 1975: Apomixis in *Xanthium*? — Watsonia **10**: 414–415.

HUMBLES, J. E., 1972: Observations on the genus *Sigesbeckia* L. — Ci. & Nat. **13**: 1–19.

JACKSON, R. C., 1960: A revision of the genus *Iva* L. — Univ. Kansas Sci. Bull. **41**: 793–876.

— 1963: Cytotaxonomy of *Helianthus ciliaris* and related species of the southwestern U.S. and Mexico. — Brittonia **15**: 260–271.

JANSEN, R. K., 1985: The systematics of *Acmella* (*Asteraceae-Heliantheae*). — Syst. Bot. Monogr. **8**: 1–115.

KATZ, M. W., TORRES, A. M., 1965: Numerical analyses of cespitose zinnias. — Brittonia **17**: 335–349.

KECK, D. D., 1960: *Calycadenia*. — In ABRAMS, L., FERRIS, R. S., (Eds.): Illustrated flora of the Pacific states 4. — San Jose: Stanford University Press.

KEIL, D. J., PINKAVA, D. J., 1976: Chromosome counts and taxonomic notes for *Compositae* from the United States and Mexico. — Am. J. Bot. **63**: 1393–1403.

— STUESSY, T. F., 1975: Chromosome counts of *Compositae* from the United States, Mexico, and Guatemala. — Rhodora **77**: 171–195.

— — 1977: Chromosome counts of *Compositae* from Mexico and the United States. — Am. J. Bot. **64**: 791–798.

LA DUKE, J. C., 1982: Flavonoid chemistry and systematics of *Tithonia* (*Compositae*). — Am. J. Bot. **69**: 784–792.

LEOPOLD, A. S., 1950: Vegetation zones of Mexico. — Ecology **31**: 507–518.

LEVIN, D. A., 1978: The origin of isolating mechanisms in flowering plants. — Evol. Biol. **11**: 185–317.

LITTLEJOHN, M. J., 1981: Reproductive isolation: a critical review. — In ATCHLEY, W. R., WOODRUFF, D. S., (Eds.): Evolution and speciation. Essays in honor of M. J. D. WHITE. — Cambridge: Cambridge University Press.

LONGPRE, E. K., 1970: The systematics of the genera *Sabazia*, *Selloa* and *Tricarpha* (*Compositae*). — Publ. Mus. Michigan State Univ., Biol. Ser. **4**: 283–384.

LÖVE, D., DANSEREAU, P., 1959: Biosystematic studies on *Xanthium:* taxonomic appraisal and ecological status. — Canad. J. Bot. **37**: 173–208.

McGREGOR, R. L., 1968: The taxonomy of the genus *Echinacea* (*Compositae*). — Univ. Kansas Sci. Bull. **48**: 113–142.

McVAUGH, R., SMITH, N. J., 1967: *Calyptocarpus vialis* and *C. wendlandii* (*Compositae*). — Brittonia **19**, 268–272.

Mears, J. A., 1973: Systematics of *Parthenium* section *Bolophytum* (*Compositae, Heliantheae*): a correlation of morphological, biochemical and habitat data. – Proc. Acad. Natl. Sci. Philadelphia **125**: 121 – 135.

– 1975: The taxonomy of *Parthenium* section *Partheniastrum* DC. (*Asteraceae-Ambrosiinae*). – Phytologia **31**: 463 – 482.

Mesfin, T., 1984: The genus *Bidens* (*Compositae*) in NE tropical Africa. – Symb. Bot. Upsal. **24**: 1 – 138.

Olorode, O., 1970: The evolutionary implications of interspecific hybridization among four species of *Zinnia* sect. *Mendezia* (*Compositae*). – Brittonia **22**: 207 – 216.

Olsen, J. S., 1979 a: Taxonomy of the *Verbesina virginica* complex (*Asteraceae*). – Sida **8**: 128 – 134.

– 1979 b: Systematics of *Zaluzania* (*Asteraceae: Heliantheae*). – Rhodora **81**: 449 – 501.

Ornduff, R., 1969: Reproductive biology in relation to systematics. – Taxon **18**: 121 – 133.

Panero, J. L., 1986: The systematics of *Viguiera* section *Eriophorae*. – Masters Thesis. Knoxville: University of Tennessee.

Parker, E. S., Jones, S. B., 1975: A systematic study of the genus *Balduina* (*Compositae, Heliantheae*). – Brittonia **27**: 355 – 361.

Parker, K. F., 1962: The South American species of *Hymenoxys* (*Compositae*). – Leafl. W. Bot. **9**: 197 – 224.

Parks, J. C., 1973: A revision of North American and Caribbean *Melanthera* (*Compositae*). – Rhodora **75**: 169 – 210.

Payne, W. W., 1964: A re-evaluation of the genus *Ambrosia* (*Compositae*). – J. Arnold Arbor. **45**: 401 – 438.

– 1966: Notes on the ragweeds of South America with the description of two new species: *Ambrosia pannosa* and *A. parvifolia* (*Compositae*). – Brittonia **18**: 28 – 37.

Peterson, K. M., Payne, W. W., 1973: The genus *Hymenoclea* (*Compositae: Ambrosieae*). – Brittonia **25**: 243 – 256.

Pinkava, D. J., 1967: Biosystematic study of *Berlandiera* (*Compositae*). – Brittonia **19**: 285 – 298.

– Keil, D. J., 1977: Chromosome counts of *Compositae* from the United States and Mexico. – Am. J. Bot. **64**: 680 – 686.

Powell, A M., 1965: Taxonomy of *Tridax* (*Compositae*). – Brittonia **17**: 47 – 96.

– 1968: Chromosome numbers in *Perityle* and related genera (*Peritylanae-Compositae*). – Am. J. Bot. **55**: 820 – 828.

Powell, S. A., 1978: Chromosome numbers in *Asteraceae*. – Madroño **25**: 160 – 169.

– Sikes, S., 1970: Chromosome numbers of some Chihuahuan Desert *Compositae*. – Southw. Naturalist **15**: 175 – 186.

– Kyhos, D. W., Raven, P. H., 1974: Chromosome numbers in *Compositae* 10. – Am. J. Bot. **61**: 909 – 913.

– – – 1975: Chromosome numbers in *Compositae* 11. *Helenieae*. – Am. J. Bot. **62**: 1100 – 1103.

Rabakonandrianina, E., 1980: Infrageneric relationships and the origin of the Hawaiian endemic genus *Lipochaeta* (*Compositae*). – Pacific Sci. **34**: 29 – 39.

Robinson, H., 1979: A study of the genus *Schistocarpha* (*Heliantheae: Asteraceae*). – Smithsonian Contr. Bot. **42**: 1 – 20.

– 1981: A revision of the tribal and subtribal limits of the *Heliantheae* (*Asteraceae*). – Smithsonian Contr. Bot. **51**: 1 – 102.

Rock, H. F. L., 1957: A revision of the vernal species of *Helenium* (*Compositae*). – Rhodora **59**: 101 – 116, 128 – 158, 168 – 178, 203 – 216.

Rollins, R. C., 1950: The guayule rubber plant and its relatives. – Contr. Gray Herb. **172**: 1 – 173.

Sanders, R. W., 1977: Taxonomy of *Rumfordia* (*Asteraceae*). – Syst. Bot. **2**: 302 – 316.

SCHILLING, E. E., HEISER, C. B., 1981: Infrageneric classification of *Helianthus* (*Compositae*). − Taxon **30**: 393 − 403.

− SCHILLING, E. M., 1986: Chromosome numbers in *Viguiera* Series *Dentatae* (*Compositae*). − Syst. Bot. **11**: 51 − 55.

SEAMAN, F. C., MABRY, T. J., 1979: Sesquiterpene lactones and species relationships among the shrubby *Ambrosia* taxa. − Biochem. Syst. Ecol. **7**: 105 − 114.

SEMPLE, J. C., 1978: A revision of the genus *Borrichia* ADANS. (*Compositae*). − Ann. Missouri Bot. Gard. **65**: 681 − 693.

− SEMPLE, K. S., 1978: *Borrichia × cubana* (*B. frutescens × arborescens*): interspecific hybridization in the Florida keys. − Syst. Bot. **2**: 292 − 301.

SETTLE, W. J., FISHER, T. R., 1972: A biosystematic study of *Silphium integrifolium* MICHAUX (*Compositae*). − Ohio J. Sci. **72**: 254 − 265.

SMITH, E. B., 1974: *Coreopsis nuecensis* (*Compositae*) and a related new species from southern Texas. − Brittonia **26**: 161 − 171.

− 1976: A biosystematic survey of *Coreopsis* in eastern United States and Canada. − Sida **6**: 12 − 215.

− 1982: Phyletic trends in section *Coreopsis* of the genus *Coreopsis* (*Compositae*). − Bot. Gaz. **143**: 121 − 124.

− 1983: Phyletic trends in sections *Eublepharis* and *Calliopsis* of the genus *Coreopsis* (*Compositae*). − Am. J. Bot. **70**: 549 − 554.

SORENSEN, P. D., 1969: Revision of the genus *Dahlia* (*Compositae, Heliantheae − Coreopsidinae*). − Rhodora **71**: 309 − 365, 367 − 416.

− 1980: New taxa in the genus *Dahlia* (*Asteraceae, Heliantheae − Coreopsidinae*). − Rhodora **82**: 353 − 360.

SPEESE, B. M., BALDWIN, J. T., Jr., 1952: Chromosomes of *Hymenoxys*. − Am. J. Bot. **39**: 685 − 688.

ST. JOHN, H., 1971: The status of the genus *Wilkesia* (*Compositae*), and discovery of a second Hawaiian species. − Hawaiian Plant Studies No. 34, Occas. Pap. Bernice Pauahi Bishop Mus. **24**: 127 − 137.

STROTHER, J. L., 1976: Chromosome studies in *Compositae*. − Am. J. Bot. **63**: 247 − 250.

− 1979: Extradition of *Sanvitalia tenuis* to *Zinnia* (*Compositae − Heliantheae*). − Madroño **26**: 173 − 179.

− 1983: More chromosome studies in *Compositae*. − Am. J. Bot. **70**: 1217 − 1224.

STUESSY, T. F., 1970: The genus *Acanthospermum* (*Compositae − Heliantheae − Melampodiinae*): taxonomic changes and generic affinities. − Rhodora **72**: 106 − 109.

− 1971 a: Systematic relationships in the white-rayed species of *Melampodium* (*Compositae*). − Brittonia **23**: 177 − 190.

− 1971 b: Chromosome numbers and phylogeny in *Melampodium* (*Compositae*). − Am. J. Bot. **58**: 732 − 736.

− 1972: Revision of the genus *Melampodium* (*Compositae: Heliantheae*). − Rhodora **74**: 1 − 70, 161 − 222.

− 1973: Revision of the genus *Baltimora* (*Compositae, Heliantheae*). − Fieldiana, Bot. **36**: 31 − 50.

− 1977 a: *Heliantheae* − Systematic review. − In HEYWOOD, V. H., HARBORNE, J. B., TURNER, B. L., (Eds.): The biology and chemistry of the *Compositae*, pp. 621 − 671. − London, New York: Academic Press.

− 1977 b: Revision of *Oparanthus* (*Compositae, Heliantheae, Coreopsidinae*). − Fieldiana, Bot. **38**: 63 − 70.

− 1978: Revision of *Lagascea* (*Compositae: Heliantheae*). − Fieldiana Bot. **38**: 75 − 133.

− 1979: Cladistics of *Melampodium* (*Compositae*). − Taxon **28**: 179 − 195.

− BRUNKEN, J. N., 1979: Artificial interspecific hybridizations in *Melampodium* section *Zarabellia* (*Compositae*). − Madroño **26**: 53 − 63.

Stuessy, T. F., Crisci, J. V., 1984: Phenetics of *Melampodium* (*Compositae, Heliantheae*). — Madroño **31**: 8 – 19.

— Spooner, D. M., Evans, K. A., 1986: Adaptive significance of ray corollas in *Helianthus grosseserratus* (*Compositae*). — Am. Midl. Naturalist **115**: 191 – 197.

Sundberg, S., Cowan, C. P., Turner, B. L., 1986: Chromosome counts of Latin American *Compositae*. — Am. J. Bot. **73**: 33 – 38.

Tanowitz, B. D., 1978: *Hemizonia conjugens* (*Compositae*): distribution, chromosome number, and relationships. — Madroño **25**: 159.

— 1982: Taxonomy of *Hemizonia* sect. *Madiomeris* (*Asteraceae: Madiinae*). — Syst. Bot. **7**: 314 – 339.

Torres, A. M., 1963 a: Revision of *Tragoceras* (*Compositae*). — Brittonia **15**: 290 – 302.

— 1963 b: Taxonomy of *Zinnia*. — Brittonia **15**: 1 – 25.

— 1964 a: Revision of *Sanvitalia* (*Compositae – Heliantheae*). — Brittonia **16**: 417 – 433.

— 1964 b: Hybridization studies in perennial zinnias. — Am. J. Bot. **51**: 567 – 573.

— 1968: Revision of *Jaegeria* (*Compositae – Heliantheae*). — Brittonia **20**: 52 – 73.

— 1969: Revision of the genus *Philactis* (*Compositae*). — Brittonia **21**: 322 – 331.

Turner, B. L., 1966: Taxonomy of *Eutetras* (*Compositae – Helenieae*). — Southw. Naturalist **11**: 118 – 122.

— 1972 a: Taxonomy of *Loxothysanus* (*Compositae, Helenieae*). — Wrightia **5**: 45 – 50.

— 1972 b: Two new gypsophilous species of *Gaillardia* (*Compositae*) from northcentral Mexico. — Southw. Naturalist **17**: 181 – 190.

— 1978 a: Taxonomy of *Axiniphyllum* (*Asteraceae – Heliantheae*). — Madroño **25**: 456 – 52.

— 1978 b: A new species of *Viguiera* (*Asteraceae – Heliantheae*) from Nayarit, Mexico. — Madroño **25**: 221 – 223.

— 1980: La taxonomía del género *Aphanactis* (*Asteraceae – Heliantheae*). — Bol. Soc. Argent. Bot. **19**: 33 – 44.

— 1982 a: Taxonomy of *Neurolaena* (*Asteraceae – Heliantheae*). — Pl. Syst. Evol. **140**: 119 – 139.

— 1982 b: New taxa in *Verbesina* (sect. *Verbesinaria*) from north-central Mexico. — Southw. Naturalist **27**: 345 – 346.

— 1984: Update on the genus *Jaegeria* (*Compositae – Heliantheae*). — Phytologia **55**: 243 – 251.

— 1985: Revision of *Verbesina* sect. *Pseudomontanoa* (*Asteraceae*). — Pl. Syst. Evol. **150**: 237 – 262.

— Davies, F., 1980: *Stuessya* (*Asteraceae: Heliantheae*), a new genus from southcentral Mexico. — Brittonia **32**: 209 – 212.

— Dawson, F., 1980: Taxonomy of *Tetragonotheca* (*Asteraceae – Heliantheae*). — Sida **8**: 296 – 303.

— Morris, M. I., 1976: Systematics of *Palafoxia* (*Asteraceae. Helenieae*). — Rhodora **78**: 567 – 628.

— Powell, A. M., 1977: Taxonomy of the genus *Cymophora* (*Asteraceae: Heliantheae*). — Madroño **24**: 1 – 6.

Urbatsch, L. E., Zlotsky, A., Pruski, J. F., 1986: Revision of *Calea* sect. *Lemmatium* (*Asteraceae: Heliantheae*) from Brazil. — Syst. Bot. **11**: 501 – 514.

Van Faasen, P., 1971: Systematics of the genus *Guardiola* (*Compositae – Heliantheae*). — Ph.D. Thesis. Michigan State University, East Lansing.

Venkatesh, C. S., 1958: A cyto-genetic and evolutionary study of *Hemizonia*, section *Centromadia*. — Am. J. Bot. **45**: 77 – 84.

Weber, W. A., 1952: The genus *Helianthella* (*Compositae*). — Am. Midl. Naturalist **48**: 1 – 35.

Wells, J. R., 1965: A taxonomic study of *Polymnia* (*Compositae*). — Brittonia **17**: 144—159.
— 1969: Specific relationships between *Polymnia canadensis* and *P. laevigata* (*Compositae*). — Castanea **34**: 179—184.
Whalen, M., 1977: Taxonomy of *Bebbia* (*Compositae: Heliantheae*). — Madroño **24**: 112—123.
White, M. J. D., 1978: Modes of speciation. — San Francisco: W. H. Freeman.
Wussow, J. R., Urbatsch, L. E., 1978: A taxonomic study of the *Calea orizabensis* complex and its bearing on the nomenclature of *Verbesina standleyi* (*Asteraceae*). — Brittonia **30**: 477—482.
— — 1979: A systematic study of the genus *Tetrachyron* (*Asteraceae: Heliantheae*). — Syst. Bot. **4**: 297—318.
— — Sullivan, G. A., 1985: *Calea* (*Asteraceae*) in Mexico, Central America, and Jamaica. — Syst. Bot. **10**: 241—267.

Authors' address: Dr Scott D. Sundberg* and Dr Tod F. Stuessy, Department of Botany, Ohio State University, 1735 Neil Ave., Columbus, OH 43210, U.S.A. — *Present address: Department of Botany, University of Washington, Seattle, Washington, U.S.A.

Pl. Syst. Evol. [Suppl. 4], 99–111 (1990)

Phytoserological investigation of the tribe *Cardueae* s. l. (*Compositae*)

H. Fischer and U. Jensen

Received December 4, 1987

Key words: Angiosperms, *Compositae, Cynareae, Cardueae, Silybum marianum, Centaurea cyanus, Xeranthemum annuum, Echinops sphaerocephalus.* – Serology, legumin.

Abstract: In order to elucidate the phylogenetic relationships within the *Compositae-Cardueae* (= *Cynareae*), the serological properties of the major seed storage protein (legumin) were compared. Antisera were raised against legumins from *Silybum marianum* (*Carduinae*), *Centaurea cyanus* (*Centaureinae*), *Xeranthemum annuum* (*Carlininae*), and *Echinops sphaerocephalus* (*Echinopinae*). In Ouchterlony immunodiffusion assays the reactivities of the sera were tested using 22 cynarean species. The results strongly suggest that the traditional "*Cynareae*" concept may not reflect phylogenetic relationships and support separate tribal rank for the *Echinopinae* and *Carlininae*.

In George Bentham's (1873) system of the *Compositae* the "*Cynareae*" is one tribe out of 13 in the *Compositae*. The tribe is traditionally divided into four subtribes which are: *Carduinae, Centaureinae, Carlininae,* and *Echinopinae.* Capitula lacking true ligulate flowers, the "Cynarean" style with a thickening below its bifurcation and the occurrence of bristles on the receptacles are considered characteristic for the tribe *Cynareae*. But a thorough look at all the cynarean genera proves these typical traits to hold for the *Carduinae* and the *Centaureinae* only; the other two subtribes present exceptions in several respects. In our century, researchers have found more and more micromorphic and chemical characters which support a proposal made by Dittrich (1977), i.e., to separate *Carlininae* and *Echinopinae* from the *Cardueae* (= *Carduinae* and *Centaureinae*) and treat them as separate tribes. However, this interpretation of Bentham's "*Cynareae*" as heterogenous was not accepted by some authors (Frohne & Jensen 1979, Dahlgren 1980, Thorne 1983).

Serology in several cases (e.g., Fairbrothers & Johnson 1964: *Cornales*, Jensen 1968: *Ranunculaceae*, Kloz 1971: *Leguminosae*, Vogel 1986: *Euphorbiaceae*) has proved to be an adequate method to elucidate phylogenetic connections on the intra- and infrafamilar level. For this, one mostly compares the major seed storage protein, the legumin (Derbyshire & al. 1976). Firstly, because of its availability and its presence in most plant seeds; secondly, because (for reasons of its non-enzymatic and non-structural function) it is less subject to internal and external selection pressures which could lead to convergencies and a distorted picture of relationships between the investigated taxa. Since preliminary tests have shown the carduean seeds to be rich in protein, such an approach appeared promising in the thistle tribe.

Materials and methods

Centaurea cyanus L. (cf. "Blauer Junge"), *Silybum marianum* (L.) Gaertn., *Echinops sphaerocephalus* L., and *Xeranthemum annuum* L. were used as reference taxa: from their seeds legumin was isolated and antisera were prepared in rabbits. Additionally, crude extracts from seeds of the following species were used as antigenic agents in the serological experiments: *Arctium lappa* L., *Atractylis gummifera* L., *Carduncellus caeruleus* (L.) DC., *Carduus nutans* L., *Carlina vulgaris* L., *Carthamus tinctorius* L., *Cirsium arvense* (L.) Scop., *Cnicus benedictus* L., *Crupina vulgaris* Cass., *Cynara scolymus* L., *Jurinea cyanoides* DC., *Lactuca serriola* L., *Lamyropsis cynaroides* (Lam.) Dittrich, *Leuzea centauroides* (L.) Holub, *Mantisalca salmantica* (L.) Briq. & Cavill. (only in reference systems *Centaurea cyanus* and *Silybum marianum*), *Onopordum acanthium* L., *Picnomon acarna* (L.) Cass., *Saussurea albescens* Hook. & Thoms., *Serratula tinctoria* L.

Voucher specimens of the species investigated are deposited in the herbarium of Bayreuth University.

Legumin extraction. 10 g of seeds (including achene walls) were homogenized in a Retsch mortar mill with 0.1 M Tris-glycine buffer pH 8.2, containing 2% NaCl. The seed mash was centrifuged for 20 min at 19 000 rpm and 12 °C. The supernatant was covered by a lipid layer which was removed before defatting the solution with 1,1,2-trichlor-trifluoroethane (Freon).

The resulting crude extract was subjected to a gel filtration on Ultrogel AcA 34, the elution buffer being the same as mentioned above.

If necessary, an ion exchange chromatography was carried out to separate the legumin from other proteins present in the major protein peak from gel filtration. This column was packed with Sephadex A-50 and equilibrated with 0.1 M Tris-glycine buffer, pH 8.2, 0.05 M NaCl. Elution was performed with a linear gradient of NaCl concentration, ranging from 0.05 M to 1 M NaCl. Protein concentrations were determined photometrically.

PAGE. For rapid analyses of protein solutions, "Mighty Small" SDS-gel slabs (Hoefer) were used. A gradient from 7.5 to 20% acrylamide was established in the gels. Electrophoresis was run at pH 8.0, 4 °C, and 100 V for c. 2 h. The gels were stained with Serva Blue R and destained with methanol/acetic acid (2 : 1). To estimate the molecular weights of the protein bands, standard protein mixtures (Pharmacia) were run together with the protein samples.

Immunization. Antisera were raised in rabbits against legumin solutions which, in SDS-PAGE, proved to be relatively pure (except of *Xeranthemum*-legumin which could not be equally purified with our methods). The amount of protein per injection was normally c. 7 − 8 mg. Freund's adjuvans (incomplete) was added for the first injection.

Blood was taken from the rabbits' ear veins (c. 50 ml at a time) after an immunization cycle of four weeks (one injection a week). The collected blood was given two days at 4 °C to coagulate, then the serum was removed and frozen at − 20 °C. The serum gained after the third immunization cycle was used for the serological experiments.

Serological tests. Two-dimensional immunodiffusion assays were performed according to Ouchterlony (1947) (method described, e.g., in Roitt 1971). Gel slabs were prepared with 0.75% Bacto-Agar in 0.4 M borate buffer pH 8.0, containing 1% NaCl. The precipitation lines were protocolled after 1, 2, and 3 days. No pre-saturation experiments were carried out. The precipitation band patterns were protocolled in a 2-dimensional matrix, differentiating between spurs, double spurs, and identity reactions (no spur). For each antiserum, a serological reaction matrix was made up, arranging the tested taxa according to the degree of serological reactiviy with the antiserum.

Table 1. Reference system *Silybum marianum*. + The taxon in the column forms a spur against the taxon in the row, i.e. has a higher reactivity with the serum. − The taxon in the row forms a spur against the taxon in the column. O No spur formed, equal reactivities. x Double spur, no priority in reactivity can be given. Si *Silybum*, Ca *Carduus*, Ci *Cirsium*, Pi *Picnomon*, La *Lampyropsis*, Ar *Arctium*, Sa *Saussurea*, Le *Leuzea*, Ma *Mantisalca*, Se *Serratula*, Cn *Cnicus*, Ce *Centaurea*, Ju *Jurinea*, Cy *Cynara*, Ct *Carthamus*, Cd *Carduncellus*, Cr *Crupina*, On *Onopordum*, La *Lactuca*, Xe *Xeranthemum*, Ec *Echinops*, Cl *Carlina*, At *Atractylis*

	Si	Ca	Ci	Pi	La	Ar	Sa	Le	Ma	Se	Cn	Ce	Ju	Cy	Ct	Cd	Cr	On	La	Xe	Ec	Cl	At
Silybum (Ca)	O	O	O	O	O	+	+	+	+	+	+	+	+	+	+	+	+	+	+	+	+	+	+
Carduus (Ca)	O	O	O	O	O	+	+	+	+	+	+	+	+	+	+	+	+	+	+	+	+	+	+
Cirsium (Ca)	O	O	O	O	O	O	+	+	+	+	+	+	+	+	+	+	+	+	+	+	+	+	+
Picnomon (Ca)	O	O	O	O	O	O	O	O	O	O	O	O	+	+	+	+	+	+	+	+	+	+	+
Lamyropsis (Ca)	O	O	O	O	O	O	O	O	O	O	O	O	+	O	O	+	+	+	+	+	+	+	+
Arctium (Ca)	−	−	O	O	O	O	O	O	O	O	O	O	O	O	x	+	+	+	+	+	+	+	+
Saussurea (Ca)	−	−	−	O	O	O	O	O	+	+	+	+	O	+	+	+	+	+	+	+	+	+	+
Leuzea (Ce)	−	−	−	O	O	O	−	O	O	O	O	O	O	O	O	O	+	+	+	+	+	+	+
Mantisalca (Ce)	−	−	−	O	O	O	−	O	O	O	O	O	O	O	O	O	O	+	+	+	+	+	+
Serratula (Ce)	−	−	−	O	O	O	−	O	O	O	O	O	O	O	O	O	O	+	+	+	+	+	+
Cnicus (Ce)	−	−	−	−	O	O	−	O	O	O	O	O	O	O	O	O	+	+	+	+	+	+	+
Centaurea (Ce)	−	−	−	−	−	O	O	O	O	O	O	O	O	O	O	O	O	+	+	+	+	+	+
Jurinea (Ca)	−	−	−	−	O	O	−	O	O	O	O	x	x	x	+	+	O	+	+	+	+	+	+
Cynara (Ca)	−	−	−	−	O	O	−	O	O	O	O	O	O	O	O	O	+	+	+	+	+	+	+
Carthamus (Ce)	−	−	−	−	−	x	−	O	O	O	O	O	−	O	O	O	O	+	+	+	+	+	+
Carduncellus (Ce)	−	−	−	−	−	−	−	O	O	O	O	O	−	O	O	+	O	+	+	+	+	+	+
Crupina (Ce)	−	−	−	−	−	−	O	O	O	O	O	−	−	−	−	−	+	+	+	+	+	+	+
Onopordum (Ca)	−	−	−	−	−	−	−	O	O	O	O	−	−	−	−	−	O	O	+	+	+	+	+
Lactuca (Lactuc)	−	−	−	−	−	−	−	−	−	−	−	−	−	−	−	−	−	−	x	x	+	+	+
Xeranthemum (Cl)	−	−	−	−	−	−	−	−	−	−	−	−	−	−	−	−	−	−	−	O	x	+	+
Echinops (Ech)	−	−	−	−	−	−	−	−	−	−	−	−	−	−	−	−	−	−	−	x	O	x	+
Carlina (Cl)	−	−	−	−	−	−	−	−	−	−	−	−	−	−	−	−	−	−	−	−	x	O	+
Atractylis (Cl)	−	−	−	−	−	−	−	−	−	−	−	−	−	−	−	−	−	−	−	−	−	−	O

Table 2. Reference system *Centaurea cyanus*. Legend see Table 1

	Ce	Cd	Cn	Ct	Ma	Ju	On	Se	Le	Ca	Pi	La	Ci	Cr	Sa	Ar	Si	Cy	Ec	La	Xe	Cl	At
Centaurea (Ce)	+	+	+	+	+	+	+	+	+	+	+	+	+	+	+	+	+	+	+	+	+	+	+
Carduncellus (Ce)	–	+	○	○	○	○	+	+	+	+	+	+	+	+	+	+	+	+	+	+	+	+	+
Cnicus (Ce)	–	○	+	○	○	+	+	+	○	+	+	+	+	+	+	+	+	+	+	+	+	+	+
Carthamus (Ce)	–	○	○	+	○	○	+	+	○	+	+	+	+	+	+	+	+	+	+	+	+	+	+
Mantisalca (Ce)	–	○	○	○	+	+	+	○	○	○	○	○	○	○	+	+	+	+	+	+	+	+	+
Jurinea (Ca)	–	○	–	○	–	–	○	○	○	○	○	○	○	○	○	○	+	+	+	+	+	+	+
Onopordum (Ca)	–	–	–	–	○	–	○	+	○	○	○	○	○	+	+	+	+	+	+	+	+	+	+
Serratula (Ce)	–	–	–	–	○	○	–	–	○	○	○	○	+	+	+	+	○	○	+	+	+	+	+
Leuzea (Ce)	–	–	○	○	○	○	○	○	–	○	○	○	○	○	○	○	○	○	+	+	+	+	+
Carduus (Ca)	–	–	–	–	○	○	○	○	○	–	○	○	○	○	○	○	○	○	+	+	+	+	+
Picnomon (Ca)	–	–	–	–	○	○	○	○	○	○	–	○	○	○	○	○	○	○	+	+	+	+	+
Lamyropsis (Ca)	–	–	–	○	○	○	○	○	○	○	○	–	○	○	○	○	○	+	+	+	+	+	+
Cirsium (Ca)	–	–	–	–	○	○	–	○	○	○	○	○	–	○	○	○	+	○	+	+	+	+	+
Crupina (Ce)	–	–	–	–	○	○	–	○	○	○	○	○	○	–	○	○	+	+	+	+	+	+	+
Saussurea (Ca)	–	–	–	–	○	○	–	○	○	○	○	○	○	○	–	○	○	○	+	+	+	+	+
Arctium (Ca)	–	–	–	–	○	○	–	○	○	○	○	○	○	○	○	–	○	○	+	+	+	+	+
Silybum (Ca)	–	–	–	–	○	○	–	○	○	○	○	○	○	–	–	–	–	○	+	+	+	+	+
Cynara (Ca)	–	–	–	–	○	–	–	–	○	○	○	○	○	○	–	–	–	–	+	+	+	+	+
Echinops (Ech)	–	–	–	–	–	–	–	–	–	–	–	–	–	–	–	–	–	–	○	+	+	+	+
Lactuca (Lactuc)	–	–	–	–	–	–	–	–	–	–	–	–	–	–	–	–	–	–	–	○	+	+	+
Xeranthemum (Cl)	–	–	–	–	–	–	–	–	–	–	–	–	–	–	–	–	–	–	–	×	×	+	+
Carlina (Cl)	–	–	–	–	–	–	–	–	–	–	–	–	–	–	–	–	–	–	–	–	–	+	+
Atractylis (Cl)	–	–	–	–	–	–	–	–	–	–	–	–	–	–	–	–	–	–	–	–	–	○	○

Table 3. Reference system *Xeranthemum annuum*. Legend see Table 1

	Xe	Pi	Si	Ca	Ci	Ju	Ar	Sa	On	La	Cy	Se	Le	Ct	Cd	Cn	Ce	Cr	La	Ec	Cl	At
Xeranthemum (Cl)	−	+	+	+	+	+	+	+	+	+	+	+	+	+	+	+	+	+	+	+	+	+
Picnomon (Ca)	−	○	○	○	○	○	○	○	+	○	+	+	+	+	+	+	+	+	○	+	+	+
Silybum (Ca)	−	○	−	○	○	○	○	+	○	○	○	+	+	+	+	+	+	+	○	+	+	+
Carduus (Ca)	−	○	○	−	○	○	○	○	○	○	+	+	+	+	+	+	+	+	○	+	+	+
Cirsium (Ca)	−	○	○	○	−	○	○	○	○	○	○	+	+	+	+	+	+	+	○	+	+	+
Jurinea (Ca)	−	○	○	○	○	−	○	○	○	○	○	+	+	+	+	+	+	+	○	+	+	+
Arctium (Ca)	−	○	○	○	○	○	○	○	○	○	○	+	+	+	+	+	+	+	○	+	+	+
Saussurea (Ca)	−	○	−	○	○	○	○	○	○	○	+	+	+	+	+	+	+	+	○	+	+	+
Onopordum (Ca)	−	−	○	○	○	○	○	○	○	○	○	+	+	+	+	+	+	+	○	+	+	+
Lamyropsis (Ca)	−	○	○	○	−	○	○	−	○	−	○	○	○	○	○	○	○	+	○	+	+	+
Cynara (Ca)	−	−	○	−	−	−	−	−	−	○	−	−	○	○	○	+	○	○	−	+	+	+
Serratula (Ce)	−	−	−	−	−	−	−	−	−	○	+	○	○	○	+	○	+	○	○	+	+	+
Leuzea (Ce)	−	−	−	−	−	−	−	−	−	○	○	○	−	○	○	○	+	○	○	+	+	+
Carthamus (Ce)	−	−	−	−	−	−	−	−	−	○	○	−	○	−	○	○	○	○	○	+	+	+
Carduncellus (Ce)	−	−	−	−	−	−	−	−	−	○	○	○	○	○	○	○	○	○	○	+	+	+
Cnicus (Ce)	−	−	−	−	−	−	−	−	−	○	−	−	○	○	○	○	○	○	○	+	+	+
Centaurea (Ce)	−	−	−	−	−	−	−	−	−	○	○	○	○	○	○	○	○	○	○	+	+	+
Crupina (Ce)	−	−	−	−	−	−	−	−	−	−	○	−	−	○	−	−	○	○	−	+	+	+
Lactuca (Lactuc)	−	−	−	−	−	−	−	−	−	−	−	−	○	−	−	−	−	○	−	+	+	+
Echinops (Ech)	−	−	−	−	−	−	−	−	−	−	−	−	−	−	−	−	−	−	−	+	+	+
Carlina (Cl)	−	−	−	−	−	−	−	−	−	−	−	−	−	−	−	−	−	−	−	−	+	+
Atractylis (Cl)	−	−	−	−	−	−	−	−	−	−	−	−	−	−	−	−	−	−	−	−	○	○

Table 4. Reference system *Echinops sphaerocephalus*. Legend see Table 1

	Ec	La	Cl	At	On	Ct	Ar	Sa	Ju	Ca	Ci	Cr	Si	Cy	Pi	Le	Cn	Cd	La	Se	Ce	Xe
Echinops (Ech)		−	+	+	+	+	+	+	+	+	+	+	+	+	+	+	+	+	+	+	+	+
Lactuca (Lactuc)	−		+	+	×	×	+	×	+	×	×	+	+	+	+	+	+	+	+	+	+	+
Carlina (Cl)	−	−		○	×	×	×	×	×	×	×	×	×	+	×	×	×	×	×	×	×	+
Atractylis (Cl)	−	−	○		−	×	×	−	−	×	×	×	−	+	○	×	×	×	+	×	×	×
Onopordum (Ca)	−	×	×	+		+	○	○	○	○	○	+	+	+	○	○	+	+	+	+	+	+
Carthamus (Ce)	−	×	×	×	−		○	+	○	○	○	+	○	+	○	+	+	○	○	+	+	+
Arctium (Ca)	−	−	×	×	○	○		○	○	○	○	○	+	○	○	○	○	+	○	+	+	+
Saussurea (Ca)	−	×	×	+	○	−	○		○	○	○	+	○	○	○	○	○	○	+	+	+	+
Jurinea (Ca)	−	−	×	+	○	○	○	○		○	○	○	○	○	○	○	○	○	+	○	○	+
Carduus (Ca)	−	×	×	×	○	○	○	○	○		○	○	○	○	○	○	○	○	○	○	+	+
Cirsium (Ca)	−	×	×	×	○	○	○	○	○	○		○	○	○	○	○	○	○	○	○	○	+
Crupina (Ce)	−	−	×	×	−	−	○	−	○	○	○		+	○	○	○	○	○	○	○	+	+
Silybum (Ca)	−	−	×	+	−	○	−	○	○	○	○	○		○	○	○	○	+	○	○	○	+
Cynara (Ca)	−	−	−	×	−	−	−	○	○	○	○	○	○		○	○	○	○	○	○	+	+
Picnomon (Ca)	−	−	×	○	○	○	○	○	○	○	○	○	○	○		○	○	○	○	○	○	+
Leuzea (Ce)	−	−	×	×	○	−	○	○	○	○	○	○	○	○	○		○	○	○	○	○	+
Cnicus (Ce)	−	−	×	×	−	−	○	○	○	○	○	○	○	○	○	○		○	○	○	○	+
Carduncellus (Ce)	−	−	×	×	−	○	−	−	○	○	○	○	−	○	○	○	○		○	○	○	+
Lamyropsis (Ca)	−	−	×	+	−	○	○	−	−	○	○	○	○	○	○	○	○	○		○	○	+
Serratula (Ce)	−	−	×	×	−	−	−	−	−	○	○	○	○	○	○	○	○	○	○		○	+
Centaurea (Ce)	−	−	×	×	−	−	−	−	○	○	○	○	○	○	○	○	○	○	○	○		+
Xeranthemum (Cl)	−	−	−	×	−	−	−	−	−	−	−	−	−	−	−	−	−	−	−	−	−	

Results

General remarks. Firstly, the sera yielded more than one precipitation band in the homologous reaction, as the vaccines did not contain completely pure legumin. However, the additional band did not disturb analyzing the reactions, since the legumin precipitate was by far the strongest one. Also, the second precipitation line may be due to a second sort of legumin.

Secondly, we had to note a lot of "identity" reactions that brought about contradictions in the serological tables (two taxa give no spur with each other, but differ in some reactions with other partners). We suppose that the sera were not yet rich enough in high affinity antibodies to visualize the precipitates of all cross reactions.

The reactions of the four antisera in short. 1. *Silybum marianum (Carduinae)* (Table 1). Perhaps the most conspicuous feature of *Silybum* is the outstanding serological similarity of *Cirsium* and *Carduus* to the reference genus *Silybum*. The remaining members of the *Carduinae* and *Centaureinae* are united in a block of medium reactivity with only gradual differences (exception: *Onopordum*), while the representatives of the *Lactuceae, Echinopinae,* and *Carlininae* show significantly less similarity to *Silybum*.

2. *Centaurea cyanus (Centaureinae)* (Table 2). The *Carduinae* and *Centaureinae* taxa form again a relatively undifferentiated group, with *Onopordum* being the *Carduinae* taxon showing the most similarity to *Centaurea*. *Echinops, Lactuca,* and the *Carlininae* spp. are separated from the other taxa by weak reactivities.

3. *Xeranthemum annuum (Carlininae)* (Table 3). There are two tendencies to be observed in this reference system: Firstly, the serological separation of the *Carduinae* from the *Centaureinae*, which could not be detected as clearly with the other antisera. Secondly, the surprisingly low reactivity of the other two *Carlininae* taxa *Atractylis* and *Carlina* which are serologically remarkably different from *Xeranthemum* in spite of belonging to the same subtribe; even *Lactuca* as a member of the *Lactuceae* has a higher reactivity.

4. *Echinops sphaerocephalus (Echinopinae)* (Table 4). Of the tested taxa, *Lactuca* is obviously the most similar to *Echinops; Carlina* and *Atractylis* differ from the *Carduinae* and *Centaureinae* by yielding a lot of double spurs, while the third *Carlininae* genus, *Xeranthemum*, has a still lower reactivity.

Discussion

Carduinae. The very intense recent investigation of the achene morphology (e.g., pappus, apical plate) has led DITTRICH (1977) to the discrimination of 4 groups of genera with similar properties:

1. *Cynara, Ptilostemon, Lamyropsis, Ancathia, Notobasis, Alfredia, Olgaea*; 2. *Arctium, Cousinia, Tiarocarpus, Onopordum, Myopordon*; 3. *Modestia, Carduus, Cirsium, Silybum*; 4. *Galactites, Picnomon, Tyrimnus*; (position of *Jurinea* and *Saussurea* uncertain).

Our results doubtlessly confirm the close relationships between *Carduus, Cirsium,* and *Silybum* (DITTRICH's group 3). Although *Silybum* antisera with a still higher titer than those used might detect serological differences between the reference taxon *Silybum* and the cross reacting genera *Carduus* and *Cirsium*, these differences were not observed in our experiments. Serologically identical reactivities between a reference taxon and cross reacting taxa belonging to different genera have rarely been found and are in favour of close relationships (JENSEN & PENNER 1980). Also karyologically (number and shape of the chromosomes) these three genera are highly similar (MOORE & FRANKTON 1962). Several phytophagous insects feed on *Cirsium, Carduus,* and *Silybum,* but avoid other *Cardueae* taxa (ZWÖLFER & PATTULLO 1970, thus suggesting similar feeding stimulation substances (or the absence of repellents) in the respective genera.

Cynara holds a serologically distant position from *Carduus, Cirsium, Silybum,* and the bulk of the other tested *Carduinae;* it has a legumin that corresponds with the *Centaureinae* as well as with the *Carduinae.* This somewhat separated position of *Cynara* may be correlated with its missing laticifers which are generally present in the *Carduinae;* other exceptions are *Ptilostemon,* one *Cousinia,* and some *Saussurea* spp. (COL 1899/1901, 1904).

The position of *Saussurea* and *Jurinea* is considered uncertain within the *Carduinae* because of their unusual character combination: (a) in both genera the pappus elements are variable in length, thus resembling the *Centaureinae;* (b) *Saussurea* is karyologically different (MOORE & FRANKTON 1962) from *Cirsium, Carduus, Silybum, Onopordum* (all *Carduinae*), but also from *Cnicus* (*Centaureinae*); (c) the *Saussurea* monothiophenes are unknown elsewhere in the *Carduinae* and in the *Centaureinae,* but are also found in *Xeranthemum* (*Carlininae*); (d) *Jurinea* contains acetylene glycosides which are typical for *Serratula* (*Centaureinae*); (e) some *Jurinea* species have pubescent achenes which is atypical for the *Carduinae.*

The serological results for these two genera provide no evidence that would favor their separation from the *Carduinae,* in spite of the relatively great similarity of *Jurinea* to the *Centaureinae* (see Table 2). Especially the reactions with the anti-*Xeranthemum* serum support *Jurinea* and *Saussurea* to be "true" *Carduinae.*

The reactivity of the *Onopordum* legumin with the anti-*Silybum* serum is surprisingly low, even significantly lower than that of all tested *Centaureinae* species. *Onopordum* presents a few features uncommon in the *Carduinae:* (a) the receptacle is not chaffy but covered with ridges resembling honeycombs; (b) the pappus does not insert within a crown-like structure on the achene apex but at the longitudinal sides of the achene; (c) *Onopordum* is the only *Carduinae* containing cnicin, a terpenoid otherwise occurring in the *Centaureinae.*

As *Onopordum* is the member of the *Carduinae* being serologically closest to *Centaurea* and also to *Echinops* (while still reacting as a "true" *Carduinae* in the experiments with anti-*Xeranthemum*), we propose *Onopordum* to be a rather primitive genus in its subtribe.

The subtribe *Carduinae* as a whole is serologically well differentiated from the *Centaureinae* by a legumin that correlates better with the *Xeranthemum* legumin. The serological data thus confirm the *Carduinae* to be a natural group. However, its affinities to the *Centaureinae* must be strong because in the other three reference systems the representatives of both subtribes are not clearly separated.

Centaureinae. DITTRICH recognizes 4 groups of genera within the *Centaureinae* with reference to their fruit morphology: 1. *Serratula, Leuzea, Rhaponticum, Tricholepis, Acroptilon, Callicephalus, Karwandarina;* 2. *Centaurea, Cnicus, Carthamus, Carduncellus, Zoegea, Schischkinia;* 3. *Mantisalca, Amberboa, Goniocaulon, Russowia, Volutaria;* 4. *Crupina.*

Our serological data confirm the close connections between *Centaurea*, *Cnicus*, *Carthamus*, and *Carduncellus* and are furthermore supported by the fact that these genera share some chemical polyacetylene components (chlorhydrine, acetate) that have otherwise not been detected in this subtribe.

For *Crupina*, serologically remote from both *Centaurea* and *Silybum*, we would propose a rather advanced position within the subtribe because of its pubescent flowers which are unique in the *Centaureinae* as well as in the *Carduinae*.

The conclusions from serology for the subtribe *Centaureinae* are in accordance with what has been acknowledged by systematists since the times of BENTHAM: The *Centaureinae* are a natural group, the next relatives being the *Carduinae*. But the other two *Cardueae* subtribes, the *Carlininae* and *Echinopinae*, apparently have much looser connections to the *Centaureinae* than traditionally has been assumed.

Carlininae. The three *Carlininae* spp. tested are serologically separated from the members of the *Carduinae* and *Centaureinae*, as has been demonstrated by the reactions with the anti-*Silybum* and the anti-*Centaurea* serum. In respect to the *Echinops* reference system, *Xeranthemum* seems to differ somewhat from *Carlina* and *Atractylis*. The results from the *Xeranthemum* reference system demonstrate that there exist very few serological similarities between *Xeranthemum* on the one and *Carlina* and *Atractylis* on the other hand. *Xeranthemum* exhibits significantly more similarities to the *Carduinae* than to the *Centaureinae* (see Table 3). Thus, according to the serological data, *Xeranthemum* could possibly be derived from early *Carduinae* ancestors. This coincides with the occurrence of special monothiophenes (polyines) in *Xeranthemum* which are also found in some *Saussurea* (*Carduinae*) species.

The other two *Carlininae* spp. tested are serologically closer connected to *Echinops* than to any other of the four reference taxa. This similarity is expressed mainly in double spurs with the *Carduinae* and *Centaureinae* species. This means that *Carlina*/*Atractylis* legumins have similarities with the *Echinops* legumin, but on different sites of the molecule than the *Carduinae* and *Centaureinae legumins*. Thus, relative to *Echinops*, *Carlina* and *Atractylis* represent a different line of (legumin) evolution than the other two subtribes mentioned. Besides, there are even greater legumin similarities between *Echinops* and *Lactuca*. All that means that we must not assume very close phyletic connections of the *Carlininae* to *Echinops*.

Thus, our data contradict the generally accepted notion of a homogenous subtribe *Carlininae*. Although experimental errors should never be excluded, this can be due to (a) *Xeranthemum* being falsely included into this subtribe; (b) *Xeranthemum* representing a distinct group of taxa within the subtribe.

The first possibility would back LEONHARDT's (1949) proposal to remove the genus *Xeranthemum* and transfer it as a new subtribe *Xerantheminae* to the *Inuleae*. Serological tests with the *Inuleae* species *Buphthalmum salicifolium*, however, did not bring any support for this idea.

The second hypothesis would be consistent with some morphological and chemical features of *Xeranthemum* which are shared by *Cardopatium*, *Cousiniopsis*, *Siebera* and *Chardinia* and are not found in the rest of the *Carlininae*: the pappus consisting of scales, not of bristles; fruit detachment area lateral, not basal; lacking secretory organs; polyacetylenes occurring as thiophenes, not as furan derivatives. These genera should be tested in another investigation with an antiserum against, e.g., *Carlina*, to demonstrate whether the serological "bipartition" of the *Carlininae* can be verified also from another point of reference.

Besides these discrepancies within the *Carlininae*, the subtribe (in whatever way we circumscribe it) cannot be linked to the cynarean "core groups" *Carduinae* and *Centaureinae*. This is most evident from the reactions with the sera against *Silybum* and against *Centaurea*, where all three *Carlininae* species reveal less similarities to the reference taxon than from the *Cichorieae Lactuca*.

Echinopinae. Serologically, *Echinops* doubtlessly holds an equally isolated position within the *Cardueae* as the *Carlininae*: In the reaction with the anti-*Silybum* serum, *Echinops* exhibits a lower legumin similarity than *Lactuca*, and this is also the case with the anti-*Xeranthemum* serum and — not so clearly — with the anti-*Centaurea* serum. The other way round, the *Echinops* reference system proves *Lactuca* to be the most similar taxon tested to *Echinops*.

These serological results are supported by a lot of observations from other fields: (a) The spheroid inflorescence of the *Echinopinae* is interpreted as a cluster of uniflorous heads and unique in the *Cardueae*. (b) The straight but articulated corolla deviates from the S-shaped Cynarean corolla as well as from the unarticulated *Carlininae* corolla. (c) The *Echinopinae* achenes were covered with multicellular hairs whereas the other *Cardueae* have simple or twin hairs or no pubescence at all. (d) The pappus inserts at the longitudinal sides of the achene, not — as normally — inside a crown like structure on the apex. (e) In contrast to *Carduinae* and *Centaureinae*, *Echinops* has no secretory organs in the stem. (f) *Echinops* differs karyologically from the other *Cardueae* (MOORE & FRANKTON 1962). (g) In contrast to the other *Cardueae*, *Echinops* exhibits a cellular endosperm (PODDUBNAJA-AR-NOLDI 1931). (h) A substance typical for the *Cardueae*, the lignan glycoside arctiin, does not occur in *Echinops* (HÄNSEL & al. 1964). (i) An investigation of GOMMERS (1973) proves *Echinops* to be the only *Cardueae* genus that exhibits a significant nematicidal effect. (k) *Echinops* is not a host plant for gall-producing insects which are normally attracted by the immature capitules of *Carduinae* and *Centaureinae* plants (ZWÖLFER 1986). Since relations between plants and phytophagous animals are based on a variety of co-adaptations, the specific *Echinops* properties mentioned should have originated during a considerable time of evolution and give useful criteria for the phylogeny of both animals and plants.

Most of the special characters of *Echinops* cited are deviations from the *Carduinae/Centaureinae* pattern. Thus, we have to separate *Echinops* at least from these two subtribes. Furthermore, considering the reactions of *Echinops* legumin with the anti-*Xeranthemum* serum, and vice versa of the *Carlininae* legumins with the anti-*Echinops* serum (tribal limits marked by *Lactuca*), there is no reason to assume any close connections between *Carlininae* and *Echinopinae*.

Lactuca. *Lactuca serriola* (*Lactuceae*) has successfully been used as a "marker" for tribal limits in the *Cardueae*. Some of the typical serological reactions of *Lactuca* were also performed with other *Lactuceae* species (*Hypochoeris radicata, Leontodon autumnalis*). Therefore, these results can be regarded as representative for the whole tribe *Lactuceae*.

If the *Lactuceae* actually represented a phylogenetic line completely distinct from all other *Compositae* tribes, as has been assumed for a long time, they would most probably not take a serological position intermediate between the *Cardueae* and, e.g., the *Carlineae*. Our data, therefore, rather support the views of CARLQUIST (1976) and JEFFREY (1978) who treat the *Lactuceae* as one tribe out of several (among them: the *Cardueae*), i.e., as one evolutionary line within the *Compositae*.

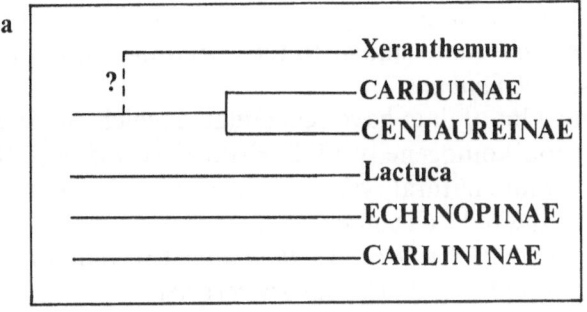

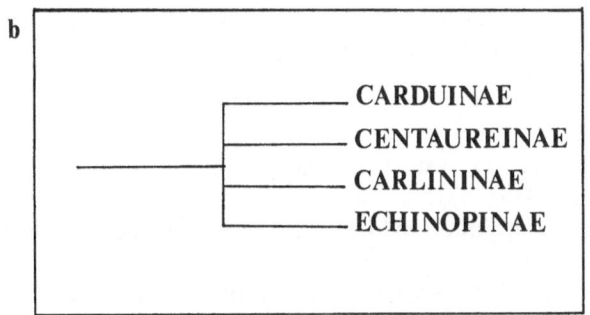

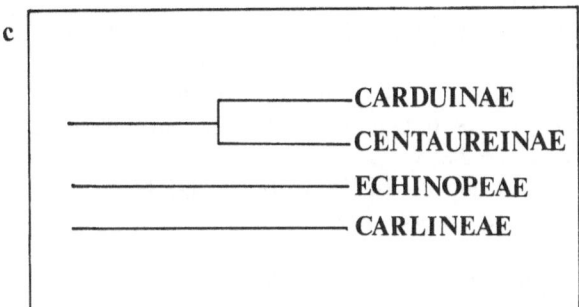

Fig. 1. *a* Possible relationships within the *Cynareae* according to serological legumin similarities; for comparison: phylogenetic outlines of the subtribes as proposed by *b* BENTHAM (1873), and *c* CASSINI (1819)

That the basic similarities between the *Cichorieae* and the rest of the *Compositae* have tended to be underestimated is also pointed out by WAGENITZ (1976).

General conclusions concerning the tribe *Cardueae* sensu BENTHAM. Serologically, the taxa included in the present study form six groups with unequal affinities to each other: On the one hand, the bulk of taxa around *Silybum* (*Carduinae*) and around *Centaurea* (*Centaureinae*), which share a lot of similarities; *Xeranthemum* has certain affinities to the *Carduinae*; the non-*Cardueae Lactuca* is relatively close to *Echinops*; finally *Carlina* and *Atractylis* (*Carlininae*) with no special affinities to either of the forementioned taxa. Figure 1 a shows the serological interrelations of the taxa investigated. For comparison, Fig. 1 b and c shows BENTHAM's (1873) and CASSINI' (1819) views of the relationships within the *Cardueae* s. l., based in either case only on morphological characters. There is conspicuous accordance between the serologically obtained *Cardueae* model (Fig. 1 a) and the ancient ideas by CASSINI (Fig. 1 c) or the scheme by DITTRICH (1977) who both treat the *Carlineae* and the *Echinopeae* as separate tribes side by side to the tribe *Cardueae* (containing the

Centaureinae and *Carduinae*). The main serological evidence in favour of this opinion is that the *Lactuceae* genus *Lactuca* is more similar to the *Cardueae* s. str. than the latter are to the *Carlininae* and *Echinopinae*.

Thus, in the course of time, quite a lot of data have accumulated which all raise severe doubts about the phylogenetic homogeneity of Bentham's *Cardueae*. It would be probably a step towards a more natural system of the *Compositae* to use the following three tribes instead of the old "*Cynareae*":

- *Cardueae* ≙ *Cynareae-Carduinae* + *Cynareae-Centaureinae* (Bentham)
- *Carlineae* (*Xeranthemum* separated?) ≙ *Carlininae* (Bentham)
- *Echinopeae* ≙ *Echinopinae* (Bentham).

We thank Heike Berthold and Anita Depser for excellent technical assistance, Dr K. Bosbach (Osnabrück) and E. Moll (Köln) for support in gaining seed materials.

References

Bentham, G., 1873: Notes on the classification, history, and geographical distribution of *Compositae*. – J. Linn. Soc. Bot. **13**: 335–577.

Carlquist, S., 1976: Tribal interrelationships and phylogeny of the *Asteraceae*. – Aliso **8**: 465–492.

Cassini, H., 1819: Sixième mémoire sur l'ordre des synanthérées, contenant les caractères des tribus. – J. Phys. Chim. Hist. Nat. Arts **88**: 152–159.

Col, A., 1899, 1901: Quelques recherches sur l'appareil sécréteur des Composées. – J. Bot. (Morot) **13**: 234–253; **15**: 166–168.

– 1904: Recherches sur l'appareil sécréteur interne des Composées. – J. Bot. (Morot) **18**: 110–133, 153–175.

Dahlgren, R., 1980: Angiospermenes taxonomi. – Kopenhagen: Akademisk Forlag.

Derbyshire, E., Wright, D. J., Boulter, D., 1976: Legumin and Vicilin, storage proteins of legume seeds. – Phytochemistry **15**: 3–24.

Dittrich, M., 1977: *Cynareae* – systematic review. – In Heywood, V. H., Harborne, J. B., (Eds.): The biology and chemistry of the *Compositae*, pp. 999–1015. – London: Academic Press.

Fairbrothers, D. E., Johnson, M. A., 1964: Comparative serological studies within the families *Cornaceae* (dogwood), and *Nyssaceae* (sour gum). – In Leone, C. A., (Ed.): Taxonomic biochemistry and serology. – New York: Ronald.

Frohne, D., Jensen, U., 1979: Systematik des Pflanzenreichs. – Stuttgart: G. Fischer.

Gommers, F. J., 1973: Nematicidal principles in *Compositae*. – Mededel. Landbouwhoogesch. Wageningen **73**: 13.

Hänsel, R. & al. 1964: Das Lignanglykosid Arctiin als chemotaxonomisches Merkmal in der Familie der *Compositae*. – Z. Naturforsch. **19 b**: 727–734.

Jeffrey, C., 1978: *Compositae*. – In Heywood, V. H., (Ed.): Flowering plants of the world. – Oxford: Elsevier.

Jensen, U., 1968: Serologische Beiträge zur Systematik der *Ranunculaceae*. – Bot. Jahrb. Syst. **88**: 204–268.

– Penner, R., 1980: Investigation of serological determinants from single storage plant proteins. – Biochem. Syst. Ecol. **8**: 161–170.

Leonhardt, R., 1949: Phylogenetisch-systematische Betrachtungen. 1. Betrachtungen zur Systematik der Compositen. – Österr. Bot. Z. **96**: 293–384.

Kloz, P., 1971: Serology of the *Leguminosae*. – In Harborne, J. B., Boulter, D., Turner, B. L., (Eds.): Chemotaxonomy of the *Leguminosae*. – London: Academic Press.

Moore, R. J., Frankton, C., 1962: Cytotaxonomic studies in the tribe *Cynareae* (*Compositae*). – Canad. J. Bot. **40**: 281–293.

OUCHTERLONY, Ö., 1947: In vitro method for testing the toxin-producing capacity of diphteria bacteria. − Abstr. 8, Scand. Pathol. Congr., Uppsala.

PODDUBNAJA-ARNOLDI, W., 1931: Ein Versuch der Anwendung der embryologischen Methode bei der Lösung einiger systematischer Fragen. 1. Vergleichende embryologisch-zytologische Untersuchungen über die Gruppe *Cynareae,* Familie *Compositae.* − Beih. Bot. Centralblatt **48**, II: 141−237.

ROITT, I. M., 1971: Essential immunology. − London: Blackwell.

THORNE, R. F., 1983: Proposed new realignments in the angiosperms. − Nordic J. Bot. **3**: 85−117.

VOGEL, CH., 1986: Phytoserologische Untersuchungen zur Systematik der *Euphorbiaceae*; Beiträge zur infrafamiliären Gliederung und zu Beziehungen im extrafamiliären Bereich. − Berlin: Dissertationes Botanicae, Cramer.

WAGENITZ, G., 1976: Systematics and phylogeny of the *Compositae.* − Pl. Syst. Evol. **125**: 29−46.

ZWÖLFER, H., 1986: Der Insektenkomplex an Disteln: ein Modell für die Selbstorganisation ökologischer Kleinsysteme. − In DRESS, A., HENDRICH, H., KÜPPERS, G., (Eds.): Selbstorganisation − die Entstehung von Ordnung in Natur und Gesellschaft. − München, Zürich: Piper.

− PATTULLO, W., 1970: Zur Lebensweise und Wirtsbindung des Distel-Blattkäfers *Lema cyanelle* L. (*Puncticollis* CURT.) (*Col., Chrysomelidae*). − Anz. Schädlingskunde, Pflanzenschutz **43**: 53−59.

Address of the author: H. FISCHER and U. JENSEN, Lehrstuhl für Pflanzenökologie und Systematik, Universität Bayreuth, Postfach 101251, D-8580 Bayreuth, Federal Republic of Germany.

Pl. Syst. Evol. [Suppl. 4], 113–119 (1990)

A geo-historical perspective on the distribution and variation in *Senecio* s. l. (*Asteraceae, Senecioneae*) in Mexico and C. America

T. M. Barkley

Received April 13, 1988

Key words: Angiosperms, *Asteraceae, Senecioneae, Senecio.* – Systematics, distribution, geohistory, plant geography. – Flora of Mesoamerica, Mexico, C. America.

Abstract: *Senecio* s. l., in Mexico and C. America includes 165 species, which occur chiefly in the highlands of C. Mexico, the Guatemalan region, Costa Rica, and adjacent Panama. The species are distributed among several convenient groups, most of which are centered in these highlands regions. A few species have affinities with temperate N. American groups or with S. American assemblages. Many species are notably intergradant with their allies, which presumably results from a fortuitous combination of a pliable and open breeding system, rapid and recent fluctuations in local climates, and the influence of mankind in creating the disturbed habitats favored by many senecios.

Preparation of the treatments of *Senecio* s. l., for the forthcoming Flora Mesoamericana and the *Asteraceae* of Mexico projects has drawn attention to the distribution patterns and the imprecise species-boundaries in the genus. Data from field and herbarium studies have been combined through traditional, intuitive techniques to recognize 165 species in the region and to cluster them into convenient species groups (Table 1). The purpose of this paper is to note that these species are largely restricted to the highlands of Mexico and C. America, and to suggest a geohistorical explanation for the observed intergradation among many of the species.

Generic and infrageneric classification in the *Senecioneae* is unsettled, and there is some enthusiasm for fracturing *Senecio* s. l., into segregate genera. For the present, however, I prefer to maintain *Senecio* as a large and variable genus, at least until a broad data base is in hand to be subjected to critical phylogenetic analyses. These matters were reviewed elsewhere (Barkley 1985 a, b) with citations of the relevant literature.

Table 1 follows the general scheme of Barkley (1985 a) and uses the same group numbers. (Groups 1–3 account for *Cacalia* s. l., and the intermediate species). The groups are informal assemblages, roughly parallel to some of the sections of Greenman (1901) and the segregate genera recognized in a series of papers by H. Robinson and colleagues (cf. Barkley 1985 a, b for citations). In

Table 1. Distribution of *Senecio* s. l. groups 4 – 11 in Mexico and C. America; groups 1 – 3 refer to *Cacalia* not treated here. Numbers indicate species in *MCA* Mexico and C. America, *CM* C. Mexico, *GUA* Guatemala Highlands, *CRP* Costa Rica-Panama Highlands; and species shared between CM-GUA and GUA-CRP, *includes spp. of Andean affinities

	MCA	CM	GUA	CRP	CM-GUA	GUA-CRP
4. *Convolvuloidei* (= *Pseudogynoxys*)	3	2	2	3	2	2
5. *Streptothamni* s. l.						
a. *Dresslerothamnus*	1			1		
b. *Nelsonianthus*	1		1			
c. *Pentacalia*	9	2	3	6	2	
6. *Pittocaulon*	5	4	1			
7. *Telanthophora*	11	7	6	1	2	1
8. *Palmatinervii* (= *Roldana*)	45	36	13	2	5	2
9. *Mulgedifolii*	11	9	3		1	
10. *Fruticosi* s. l.						
a. *Suffruticosi*	7	1				
b. *Barkleyanthus*	1	1	1		1	
c. *Fruticosi* s. str.	12	9	1	2*		
11. *Herbacei*						
a. *Multinervii*	6		1	6		1
b. *Annui*	8	1	1			
c. *Triangulares*	19	14		5*		
d. *Aureoidei*	15	8				
e. *Lugentes*	11	8	2			
Totals	**165**	**102**	**36**	**26**	**14**	**6**
Percent species	100	62	22	16	8	4
Percent land area	100	20	5	2		

the present paper, group 8 includes those "cacalioid" entities that are related to the "cacalioid" *Palmatinervii* assemblage, i.e., the segregate genus *Roldana*, but which lack the broad, palmate leaves of most of that assemblage. The disposition of these species was erroneous in BARKLEY (1985 b), where they formed a disparate element in subgroup 10 c, *Fruticosi* s. str. Addition adjustment of the taxonomic scheme is clearly called for, but that is not the focus of this paper.

The numbers of species presented in Table 1 are subject to change as the revisionary studies progress. However, the relative numbers of species for the groups and the ratios among them are not likely to be much altered.

Material

The following herbaria have lent specimens for these revisionary studies: CAS, KANU, KSC, GH, MICH, NY, MO NMC, OS, SD, TEX, UC, UMO, and the following have been consulted: B, BM, ENCB, K, MEXU, XAL.

Distribution of *Senecio*

Nearly all of the Mexican and C. American senecios occur in regions of higher elevation, above 1 800 m, which are also the regions where most of the people live at present and have done so since prehistoric times. Both field and herbarium studies indicate that many senecios are at home in disturbed or semidisturbed habitats and that many species have notably imprecise species boundaries.

C. Mexico (CM) is the region of the Trans-Mexican Volcanic Belt, the central and southern Sierra Madre Oriental, and the Sierra Madre Occidental, plus the southern Sierra Madre and associated highlands of Oaxaca and it comprises about 20% of the land area of Mexico and C. America (Fig. 1). However, the CM region includes 102 spp., or about 62% of the total for the region.The Guatemala Highlands (GUA) include southern Chiapas and adjacent Guatemala, plus small portions of El Salvador and neighboring Honduras, for about 5% of the land area of Mexico and C. America and about 22% of the senecios. The Highlands of Costa Rica and western Panama (CRP) account for about 2% of the land area and 16% of the senecios. There are obviously senecios elsewhere in Mexico and C. America outside of the highland regions, but most of those species also occur in the highlands. Nicaragua lacks extensive highlands and has few senecios; nearly all of them are species that also occur in the Guatemala highlands to the northwest. There are

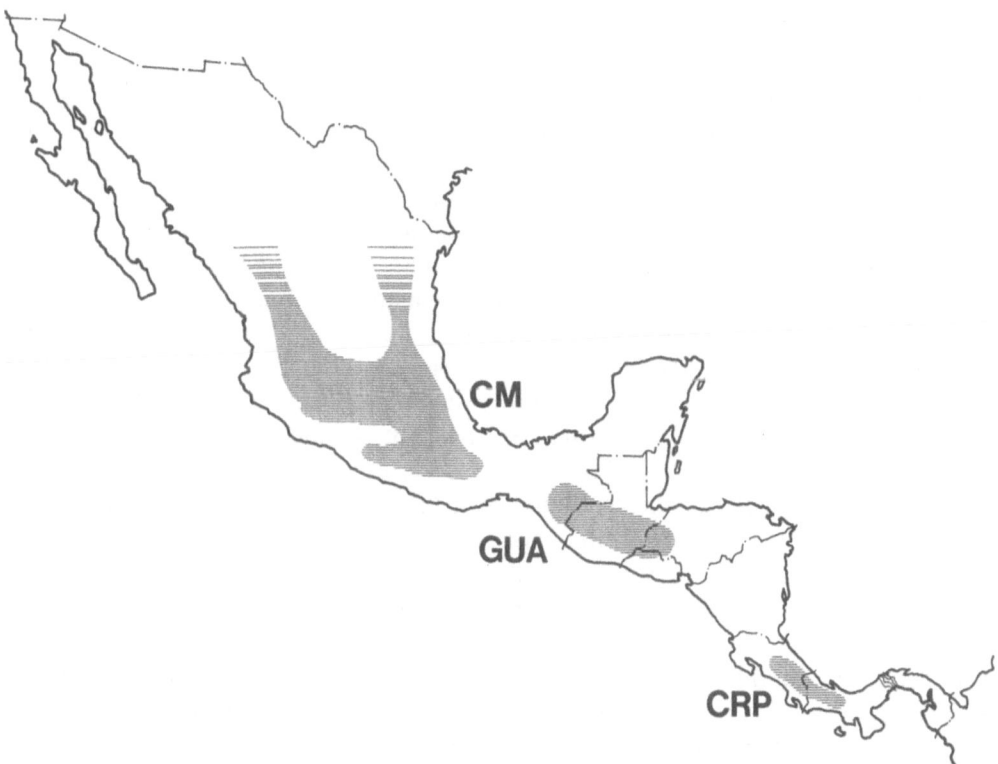

Fig. 1. Highland regions in Mexico and C. America; CM C. Mexico, GUA Guatemala Highlands, CRP Highlands of Costa Rica and Panama

also some species in the north and west of Mexico that are not affiliated with those of the highlands flora.

It is noteworthy that most of the senecios of the Guatemala Highlands and of the Costa Rica-Panama Highlands are comfortably disposed among those groups that are primarily or exclusively developed in the C. Mexican region. The exceptions are rather few. Group 4 (*Convolvuloidei*, = the segregate genus *Pseudogynoxys*) has three species in Mexico and C. America, but is well represented in S. America (ROBINSON & CUATRECASAS 1977). The distribution of this group is complicated because the plants are attractive, climbing vines, and they have been widely scattered through intentional cultivation. The nine species referable to subgroup 5 b (*Pentacalia*) are related to a large Andean complex (ROBINSON & CUATRECASAS 1978). The species of subgroups 10 a (*Suffruticosi*) and 11 d (*Aureoidei*) in Mexico are merely the southern ends of more diverse distributions in temperate N. America (BARKLEY 1978, FREEMAN 1985). Group 11 b (*Annui*) is a mixed assemblage that includes some desert natives of northern Mexico, as well as the widespread Old World weed, *Senecio vulgaris* L. Finally, there are two small shrubs and four coarse herbs of apparent Andean affinities that occur in the distinctive paramo of high elevations in Costa Rica. These plants of the paramo are included among subgroups 10 c (*Fruticosi* s. str.) and 11 c (*Triangulaces*) in Table 1, purely for convenience.

All of the species of groups 6 to 9, plus most of the species in groups 10 and 11, belong to assemblages that are largely restricted to the highlands of Mexico and C. America, and show their greatest diversity there.

Species delimitation

There is an abundant literature documenting the imprecision of many species boundaries for *Senecio* in N. America and elsewhere (BAIN 1983; BARKLEY 1962, 1968 a, b, 1980, 1988; BRUNERYE 1969; KOWAL 1975; VILLASEÑOR 1986), and it is no surprise that many senecios in Mexico and C. America are complicated by apparent intergradation. Field and herbarium studies indicate that most assemblages include some species that are widespread and intergradant with each other, along with some species that are more or less distinctive but variable, plus some species that are both geographically restricted and morphologically distinctive. The widespread and variable species are distinguished on the basis of perceived clusters and characters that show nearly continuous variation, i.e., the species are separated by low points in a "curve" depicting variation. In practice, most of the relevant specimens are referable to a "species", but a certain portion, perhaps a fifth of them, are seen as variously intermediate. Continuous variation and the incumbent taxonomic problems are facts of life among many senecios, and the present sources of data suggest no satisfacory resolution.

Variation among species is confined to those species within each group or subgroup, except for limited intergradation between some species currently assigned to subgroups 10 c and 11 c, thus suggesting the artificialy of these groups. Moreover, the intergradation is greatest and the species are least precise in the highlands of C. Mexico, where the numbers of species and the land area are greatest.

The question arises as to why so many species of *Senecio* are intergradant and poorly defined. A speculative but reasonable explanation incorporates the biological facts of the group with the geohistory of Mexico and C. America.

There is no direct information on the breeding mechanisms of the senecios in the Mexican and C. American highlands, and scant information from elsewhere. The data for the aureoid assemblage were reviewed elsewhere (BARKLEY 1988), with the conclusion that the species of that group are largely obligate outbreeders, with low barriers to hybridization, and with introgression as a frequent possibility. There is a body of experimental studies on the Old World weedy species, *Senecio vulgaris* L. and *Senecio squalidus* L., which shows, among other things, that hybridization and gene flow between these species are responsible for certain morphological variation (cf. Ross & ABBOTT 1987, for citations and discussion). Field observations in C. Mexico indicate that senecios in flower are normally visited by an abundance of flying insects, and presumably, pollination is by generalist vectors. The cytology of the Mexican and C. American entities is based upon n = 20, or n = 30, or numbers derived from either of these two base numbers (ROBINSON & BRETTELL 1974, NORDENSTAM 1977, FREEMAN 1985) leading to the easy conclusion that the species may be historically of polyploid origin, but they are now functioning as diploids (BARKLEY 1988). In light of the foregoing and the absence of contrary evidence, it is reasonable to assume that these variable and intergradant senecios are normally outcrossers, with generalized pollinators, and that the cytological structure allows easy hybridization and introgression.

The geohistory of Mexico and C. America is complicated and poorly understood, with scattered data and few syntheses. Among the more recent syntheses are those of BROWN & GIBSON 1983, CONEY (1982), GENTRY (1982), GRAHAM (1973), RZEDOWSKI (1978), and TOLEDO (1982). A digestion of these secondary sources, plus personal communication with Alan GRAHAM, lead to the following generalities. The *Asteraceae* were present in the Mesoamerican and Mexican flora by the late Tertiary, and presumably senecios or their immediate progenitors, were among them. Through the late Tertiary and the Quaternary and to the present, the region has been one of great tectonic activity, which is reflected in extensive volcanism, mountain building, and climatic changes. Apparently, warm-cool and wet-dry climatic fluctuations have varied independently of each other throughout the timespan, creating a continuing abundance of different and changing habitats in rather close proximity to each other.

A plausible scenario is that the antecedants of the present senecios were diversified into the Mexican and C. American highlands regions in the late Tertiary. As the geotectonic events and climatic variations created new niches, the senecios were able to radiate into them, as a result of their easy, open breeding system and ability to hybridize freely. Presumably, the persistent pappus of the mature achenes enhanced their opportunities for dispersal. The ability of many senecios to thrive in disturbed or semidisturbed habitats likewise enhanced their chances of becoming abundant and widespread. Eventually, the continuing climatic changes produced a reticulating pattern of short-term isolation and semispeciation, followed by reintegration of populations, thus giving rise to the great abundance of weakly separated entities that are seen today. These entities of Mexico and C. America, and most especially those of groups 8, 10, and 11, behave in a manner similar to the aureoids of temperate N. America (BARKLEY 1988).

A complicating factor, unique to the Mexican and C. American highlands, is the role of mankind in the region. Sedentary agriculture there is ancient; irrigated farming has been practiced in the Tehuacan Valley of C. Mexico for some 6 000 years (MacNeish 1972 a, b). These, plus other agricultural practices, have certainly added to the disturbed habitats favored by many widespread senecios. It is, therefore, possible that the long-time habitat alterations by mankind also have been a significant factor in creating the diversity among the senecios of Mexico and C. America.

I am grateful to the administrations of the herbaria and to the following individuals for advice and assistance: John Bain, Arthur Cronquist, Alan Graham, T. P. Ramamoorthy, Barbara Ralston, Harold Robinson, Eileen Schofield, Tod Stuessy, B. L. Turner, and Jose Luis Villaseñor.This paper is contribution no. 88-434-J from the Kansas Agricultural Experiment Station, Manhattan.

References

Bain, J., 1983: A biosystematic study of the *Senecio streptanthifolius* complex. – Doctoral dissertation. Edmonton: The University of Alberta.

Barkley, T. M., 1962: A revision of *Senecio aureus* and allied species. – Trans. Kansas Acad. Sci. **64**: 318–408.

– 1968 a: Intergradation of *Senecio* sections *Aurei, Tomentosi* and *Lobati* through *Senecio mutabilis*. – Southwest. Naturalist **13**: 109–115.

– 1968 b: Taxonomy of *Senecio multilobatus* and its allies. – Brittonia **20**: 267–284.

– 1978: *Senecio*. – North Am. Fl. II. **10**: 50–139.

– 1980: Taxonomic notes on *Senecio tomentosus* and its allies. – Brittonia **32**: 281–308.

– 1985 a: Generic boundaries in *Senecioneae*. – Taxon **34**: 17–21.

– 1985 b: Infrageneric groups in *Senecio* s. l., and *Cacalia* s. l. (*Asteraceae: Senecioneae*) in Mexico and Central America. – Brittonia **37**: 211–218.

– 1988: Variation among aureoid senecios of North America: a geohistorical interpretation. – Bot. Rev. **54**: 82–106.

Brown, J. H., Gibson, A. C., 1983: Biogeography. – St. Louis: C. V. Mosby.

Brunerye, L., 1969: Les *Senecio* du groupe *Helenitis*. – Doctoral dissertation. Rouen: Université de Rouen.

Coney, P. J., 1982: Plate tectonic constraints on the biogeography of Middle America and the Cordilleran region. – Ann. Missouri Bot. Gard. **69**: 432–443.

Freeman, C. C., 1985: A revision of the aureoid species of *Senecio* (*Asteraceae: Senecioneae*) in Mexico, with a cytogeographic and phylogenetic interpretation of the aureoid complex. – Doctoral dissertation. Manhattan: Kansas State University.

Gentry, A. H., 1982: Neotropical floristic diversity. – Ann. Missouri Bot. Gard. **69**: 557–593.

Graham, A., (Ed.), 1973: Vegetation and vegetational history of northern Latin America. – Amsterdam: Elsevier.

Greenman, J. M., 1901: Monographie der nord- und zentralamerikanischen Arten der Gattung *Senecio*. 1. Allgemeines und Morphologie. – Leipzig. (Reprinted in Englers Bot. Jahrb. Syst. **32**: 1–33. 1902.)

Kowal, R. R., 1975: Systematics of *Senecio aureus* and allied species in the Gaspé Peninsula, Quebec. – Mem. Torrey Bot. Club **23**: 1–113.

MacNeish, R., (Ed.), 1972 a: The prehistory of the Tehuacan Valley, 4, chronology and irrigation. – Austin: University of Texas Press.

– 1972 b: The evolution of community patterns in the Tehuacan Valley of Mexico and speculations about cultural processes. – In Ucko, P. J., & al., (Eds.): Man, settlement and urbanism. – Cambridge: Schenkman.

Nordenstam, B., 1977: *Senecioneae* and *Liabeae* – systematic review. – In Heywood, V. H. & al., (Eds.): The biology and chemistry of the *Compositae* 2, pp. 799–830. – London: Academic Press.

ROBINSON, H., BRETTELL, R. D., 1974: Studies in the *Senecioneae* (*Asteraceae*). 5. The genera *Psacaliopsis, Barkleyanthus, Telanthophora*, and *Roldana.* – Phytologia **27**: 402 – 439.

– CUATRECASAS, J., 1977: Notes on the genus and species limits of *Pseudogynoxys* (GREENM.) CABRERA (*Senecioneae, Asteraceae*). – Phytologia **36**: 177 – 192.

– – 1978: A review of the Central American species of *Pentacalia* (*Asteraceae: Senecioneae*). – Phytologia **40**: 37 – 50.

ROSS, M. D., ABBOTT, R. J.,1987: Fitness, sexual asymmetry, functional sex and selfing in *Senecio vulgaris* L. – Evol. Trends in Plants **1**: 21 – 28.

RZEDOWSKI, J., 1978: Vegetación de México. – México: Limusa.

TOLEDO, V. M., 1982: Pleistocene changes in vegetation in tropical Mexico. – In PRANCE, G. T., (Ed.): Biological diversification in the tropics. – New York: Columbia University Press.

VILLASEÑOR, J. L., 1986: Revisión de *Senecio* sección *Mulgedifolii* (*Compositae: Senecioneae*). – Master's thesis. México: Universidad Nacional Autónoma de México.

Address of the author: T. M. BARKLEY, Herbarium, Division of Biology, Kansas State University, Manhattan, KS 66506, U.S.A.

Index to plant names (tribes and genera)

M. Hesse · F. Ehrendorfer (eds.)
Morphology, Development, and Systematic Relevance of Pollen and Spores

1990. 122 figures. VII, 124 pages.
Cloth DM 138,–, öS 980,–
Reduced price for subscribers to "Plant Systematics and Evolution":
Cloth DM 124,20, öS 882,–
ISBN 3-211-82182-1

Palynology (the science of fossil and recent spores/pollen grains) is of high importance, both in many pure and applied fields of the natural sciences (e.g. in botany, geology, climatology, archeology, and medicine). It is not only an auxilliary science, but certainly can stand for itself. The "classical" palynology subjects, pollen morphology and systematics, are at present influenced by many modern approaches, e.g. from cell biology, analytical electron microscopy, morphometry, up to computer-aided-design of threedimensional reconstruction. In recent years fascinating informations have come to light, and new insights have given rise to changing scientific concepts. During the XIV International Botanical Congress, held in Berlin in 1987, a symposium was devoted to important topics of (actuo)palynology. Nine of its innovative, major contributions are presented in this volume. They cover the comparative morphology and the systematic/evolutionary significance of pollen/spores in critical taxa, aspects of pollen development (cytoskeleton), the substructure of sporopollenin, homologies between wall strata of ferns, gymnosperms and angiosperms, and important (but so far underrated) physical aspects of harmomegathy and pollen transport (fluid versus solid mechanics).

Plant Systematics and Evolution · Supplementum 5

Springer-Verlag Wien New York

Springer-Verlag, Mölkerbastei 5, P.O. Box 367, A-1011 Wien · Heidelberger Platz 3, D-1000 Berlin 33· 175 Fifth Avenue, New York, NY 10010, U.S.A. · 37-3, Hongo 3-chome, Bunkyo-ku, Tokyo 113, Japan.